# The Biopolitics of Lifestyle

This work focuses on the biopolitical use of lifestyle to govern individual choice and secure population health from the threat of obesity. The characterization of obesity as a threat to society, caused by the cumulative effect of individual lifestyles, has led to the politicization of daily choices, habits and practices as potential threats. This book critically examines these unquestioned assumptions about obesity and lifestyle, and their relation to wider debates surrounding neoliberal governmentality, biopolitical regulation of populations, discipline of bodies, and the possibility of community resistance.

The rationale for this book follows Michel Foucault's approach of problematization, addressing the way lifestyle is problematized as a biopolitical domain in neoliberal societies. Mayes argues that in response to the threat of obesity, lifestyle has emerged as a network of disparate knowledge, relations and practices through which individuals are governed toward the security of the population's health. Although a central focus is government health campaigns, this volume demonstrates that the network of lifestyle emanates from a variety of overlapping domains and disciplines, including public health, clinical medicine, media, entertainment, school programmes, advertising, sociology and ethics.

This book offers a timely critique of the continued interventions into the lives of individuals and communities by government agencies, private industries, medical and non-medical experts in the name of health and population security and will be of interest to students and scholars of critical international relations theory, health and bioethics, and governmentality studies.

**Christopher Mayes** is a Postdoctoral Research Fellow at the Centre for Values, Ethics and the Law in Medicine (VELiM), University of Sydney.

# Interventions

Edited by Jenny Edkins, Aberystwyth University and
Nick Vaughan-Williams, University of Warwick

The Series provides a forum for innovative and interdisciplinary work that engages with alternative critical, post-structural, feminist, postcolonial, psychoanalytic and cultural approaches to international relations and global politics. In our first 5 years we have published 60 volumes.

We aim to advance understanding of the key areas in which scholars working within broad critical post-structural traditions have chosen to make their interventions, and to present innovative analyses of important topics. Titles in the series engage with critical thinkers in philosophy, sociology, politics and other disciplines and provide situated historical, empirical and textual studies in international politics.

We are very happy to discuss your ideas at any stage of the project: just contact us for advice or proposal guidelines. Proposals should be submitted directly to the Series Editors: Jenny Edkins (jennyedkins@hotmail.com) or Nick Vaughan-Williams (n.vaughan-williams@warwick.ac.uk).

'As Michel Foucault has famously stated, "knowledge is not made for understanding; it is made for cutting". In this spirit the Edkins–Vaughan-Williams Interventions series solicits cutting-edge, critical works that challenge mainstream understandings in international relations. It is the best place to contribute post-disciplinary works that think rather than merely recognize and affirm the world recycled in IR's traditional geopolitical imaginary.'

*Michael J. Shapiro, University of Hawai'i at Manoa, USA*

# The Biopolitics of Lifestyle

Foucault, ethics and healthy choices

**Christopher Mayes**

Routledge
Taylor & Francis Group

LONDON AND NEW YORK

First published 2016 by Routledge

2 Park Square, Milton Park, Abingdon, Oxfordshire OX14 4RN
711 Third Avenue, New York, NY 10017

*Routledge is an imprint of the Taylor & Francis Group, an informa business*

First issued in paperback 2017

*British Library Cataloguing in Publication Data*
A catalogue record for this book is available from the British Library

Library of Congress Cataloging in Publication Data
A catalog record for this book has been requested

ISBN: 978-1-138-93386-6 (hbk)
ISBN: 978-0-8153-7739-9 (pbk)

Typeset in Times New Roman
by HWA Text and Data Management, London

# Contents

# Acknowledgements

The research, writing and re-writing of this book has occurred across different continents, different institutions and with different interlocutors. To these people and circumstances I owe a great deal.

The initial iteration of this research was a dissertation under the supervision of Catherine Mills at the Centre for Values, Ethics and the Law in Medicine, University of Sydney. Her critical guidance and continued engagement has been invaluable in developing the ideas and arguments presented in this book. Sam Binkley, Cressida Heyes and Lars Thorup Larsen also provided substantial feedback on the dissertation. The close readings and insightful suggestions offered through their examiners' reports were significant in transforming the dissertation into a book.

I held a Postdoctoral Fellowship at the Rock Ethics Institute, Pennsylvania State University from 2011 to 2013. This presented the opportunity to work with Don Thompson and Jonathan Marks on the ethics of food and institutional corruption. The many long conversations with Don and Jonathan about ethics, biopolitics and public health governance deeply influenced the re-drafting and re-writing process. I am also grateful to Nancy Tuana and the Rock Ethics Institute for allowing time to work on this project.

Over the past eight years I have received continual support from colleagues at the Centre for Values, Ethics and the Law in Medicine, University of Sydney. Ian Kerridge and Stacy Carter have always been ready to offer advice and encouragement in the development of my research. The friendship of Scott Fitzpatrick and Peter Lewis has been an important source of care and support. I am especially grateful to Jane Williams and Jenny Kaldor for reading draft chapters and providing valuable recommendations.

The anonymous reviewers and the editorial team at Routledge have also been instrumental in shaping this book and refining its argument.

In addition to the academic environment, my family has been a profound source of inspiration and their support has always been felt. To my wife Eve, I owe the greatest praise and thanks. Her commitment to reading drafts, debating ideas over dinner, calling me out for talking crap, encouraging me to write and patiently listening to my rants has been a blessing and a joy.

Chapter 1 and Chapter 6 develop ideas initially discussed in Mayes, C. (in-press). "Revisiting Foucault's 'Normative Confusions': Surveying the debate since the Collège de France lectures." *Philosophy Compass*. (Wiley-Blackwell).

Chapter 1 draws on Mayes, C. (2015). "The Harm of Bioethics: A Critique of Singer and Callahan on Obesity." *Bioethics* 29(3): 217–221 (Wiley-Blackwell), and Mayes, C. (2010). "The Violence of Care: An Analysis of Foucault's Pastor." *Journal of Cultural and Religious Theory* 11(1): 111–126.

Chapter 2 includes extracts from Mayes, C. (2014). "Governing through Choice: Food Labels and the Confluence of Food Industry and Public Health to Create 'Healthy Consumers'." *Social Theory and Health* 12, no. 4. (Palgrave Macmillan).

# Introduction

## The lifestyle problematic

Since 1997 the World Health Organization has been raising the alarm over obesity, arguing that it is a 'growing threat to health in countries all over the world' (World Health Organization 2000). Ordinarily occupied by concerns over malnutrition and outbreaks of communicable diseases, the WHO now fears the global effects of bodies, eating habits and daily activity of men, women and children. National governments, public health authorities and diverse health-related agencies also express concern over the rising incidence of obesity. Child protection agencies suggest that childhood obesity can be a form of parental abuse warranting the removal of children and/or requiring them to undergo bariatric surgery (Murtagh and Ludwig 2011, Lowe 2012). Paediatricians call for programmes that track and measure the weight of school children (Hagan 2012, Lacy et al. 2012) and monitor infant feeding practices through domestic visitations (Wen et al. 2011). It is not only health and medical professionals recommending interventions. Economists and statisticians calculate that obese people should pay more for health insurance and transportation due to their weight, size and purported risk of developing chronic disease (Roberts and Edwards 2010, Leonhardt 2009). And ethicists argue that obesity is not a private matter but an ethical issue due to the impositions that obese people place on others (Singer 2012). It is believed that these impositions are so great that governments should deploy 'coercive public health measures' to pressure overweight and obese people to make healthy choices (Callahan 2013).

Responding to these concerns, governments in the United States, Australia, the United Kingdom and elsewhere have launched a variety of social marketing campaigns and community programmes encouraging individuals to monitor diet and exercise behaviours. A striking, yet common theme among national campaigns is a fear that obesity threatens the security of the population. In the US, Michelle Obama launched *Let's Move* because she worries that '[t]he physical and emotional health of an entire generation and the economic health and security of our nation is at stake' (Office of the First Lady 2010). The Australian Government fears obesity as 'one of the greatest public health challenges' (National Preventative Health Taskforce 2009: 1). And the UK Government laments that '[a]t a time when our country needs to rebuild our economy, overweight and obesity impair the productivity of individuals and increase absenteeism' (Department of Health 2011a). With almost unanimous social and political agreement that something *must* be done, obesity proves to be a unifying issue that dissolves party lines and mobilizes social workers, paediatricians, economists, dieticians, social marketers, ethicists and others into action.

The near universal agreement on obesity as a problem requiring urgent action is largely due to the way this problem, and its solution, is articulated. Individual choices and behaviours are the problem and the solution is for individuals to make responsible choices and adopt healthy behaviours. That is, individual lifestyles are the cause of potential public health, economic and nation security catastrophes, but also the cure. Andrew Lansley, former UK Secretary of State for Health, effectively summarizes the political rhetoric used in many Western liberal-democracies in stating, '[w]e are now all too familiar with the impact of modern lifestyles on the health and wellbeing of the population, and the consequences for our economy' (Department of Health 2011). The prevailing governmental rationality is that individual lifestyles are both to blame and the source of salvation in relation to obesity. Framed in this way, the role for government and industry is simply to create an environment in which individuals can make 'healthy lifestyle choices'.

The attempt by governments to modify the health-related behaviours of individuals often provokes criticism from groups concerned with individual freedom and government overreach. Since the late 1970s a small group of scholars have critiqued lifestyle-focused health campaigns under the banner of 'healthism'. Petr Skrabanek described healthism in Western democracies as when 'the state goes beyond education and information on matters of health and uses propaganda and various forms of coercion to establish norms of a "healthy lifestyle" for all' (Skrabanek 1994: 15ff). Aspects of my analysis overlap with the healthism critiques of the 1980s and 90s (Crawford 1980, Fitzpatrick 2001, Fitzgerald 1994), particularly the exposure of ideological underpinnings that divide human activity 'into approved and disapproved, healthy and unhealthy, prescribed and proscribed, responsible and irresponsible' (Skrabanek 1994). However, the overreliance on libertarian notions of individual freedom and hypersensitivity to any public health guidance diminishes the analytic utility of the healthism critique. The healthism literature is peppered with tropes such as 'tyranny', 'coercion', 'nemesis', 'propaganda' and 'unfreedom' to describe public health (Fitzpatrick 2001, Illich 1975). These concerns arguably reflect a Cold War climate that was suspicious of state interference. While the concept of healthism can be useful, I contend that a more sensitive set of analytic and conceptual tools are needed to address a different set of questions that deal with the role of lifestyle in political disputes over individual freedom, population security, and the governance of the future.

The focus of this book is on the use of lifestyle as a governmental mechanism or device. Although different disciplines attempt to define lifestyle as stable concept, I argue that it is best thought of as a network of disparate ideas, beliefs and practices through which individual choices and bodies are governed. For instance, public health and epidemiology define lifestyle as a set of habits (diet and exercise) and substance consumption (tea, coffee, tobacco or alcohol) that are associated with health status (Porta 2008: 143). Sociology uses lifestyle to refer to class distinction and status, to differentiate rural and urban forms of life and to examine consumption and identity in consumer societies (Abercrombie, Hill, and Turner 2006: 222). Political science considers lifestyle as the choices that reflect political attitudes and goods (Bealey 1999: 193). To understand the current political uses of lifestyle it is necessary to examine the way these different disciplines combine and overlap to produce knowledge that contributes to the development of lifestyle as mechanism of governance, particularly in the context of obesity. My intention is not to uncover what lifestyle *is* but to trace the multi-linear lines of knowledge, entangled relations of power and enmeshed practices of the self that comprise the network of lifestyle and enable the governance of individuals.

Lifestyle is used as a governmental mechanism in a variety of areas, however in recent years it has become ubiquitous in obesity discourse. This book uses the rhetoric and policies surrounding the obesity epidemic as the backdrop for an investigation into lifestyle as a biopolitical contrivance. The overlapping conceptions of lifestyle are highlighted by the entangled discourses and strategies targeting obesity. It is not only national governments that are concerned about the link between obesity and lifestyle. Government campaigns and rhetoric are influential in shaping norms of health and the body, however the media in all its guises, food and diet industries, economists and fund managers, dieticians, personal trainers, researchers and a plethora of others contribute to the notion of lifestyle and its relationship to obesity. The diversity of knowledge and guidance reinforce lay perceptions that a clear causal relationship exists between individual choices and the obesity epidemic. As such, an expectation is established that rational individuals should choose in a manner that promotes their own health and does not burden the population. The widespread acceptance of obesity as a threat arising through individual choice provides the conditions for interventions into the everyday choices and activities of individuals, communities and the population.

At first blush, the focus on individual lifestyle appears straightforward and justified. If individual choices and lifestyles are causing 'one of the greatest public health challenges' that threatens the security of 'industrialized countries' (National Preventative Health Taskforce 2009: 1), then it is appropriate and necessary to target the individual's everyday life. Yet, there are a significant number of detractors from this logic that reduces the obesity epidemic to individual choices. At one end of the spectrum are critics who argue that obesity is a social construction. Proponents of this position argue that the obesity is a 'myth' supported by 'junk science' that is skewed by 'crass economic motives' (Campos 2004: xix), resulting in the medicalization of diversity and the normalization of bodies (Murray 2008: 7, Wann 2009: xiii). In the middle of the spectrum are a growing minority of public health researchers and epidemiologists that acknowledge the existence of obesity at the individual level but suggest the notion of an epidemic threatening the security of the population relies on inadequate epidemiological data (Gard 2011: 17, Flegal et al. 2013). At the other end of the spectrum are researchers that accept the obesity epidemic as a significant public health problem. However, they argue that a focus on the individual is ethically questionable, ineffective and ignores relevant research that suggests systemic factors, not individual choices, are behind the epidemic (Baum 2011, Brownell et al. 2010, Marmot 2000, Gordon-Larsen et al. 2006, Howell and Ingham 2001, Carter et al. 2011, Lang and Rayner 2010). This spectrum of critics, which could be further refined, raises doubts about the appropriateness of anti-obesity campaigns targeting individual choice and lifestyle. At the very least, a growing body of research disputes the evidence base of lifestyle-focused anti-obesity measures.

Drawing on the work of Michel Foucault, I trace the emergence of lifestyle through the question: how did lifestyle come to be conceived as a domain of biopolitical experience? Or put differently, how did the everyday life, choices and activities of the individual come to be politicized as capable of influencing the security and health of the population and therefore requiring governance through an idea of lifestyle? Through exploring these questions I argue that the ambiguity of lifestyle, in combination with the ubiquity of its mobilization, presents a problem both for how it is used and how it developed. In questioning and examining the development of lifestyle, particularly its use in normalizing health and bodies, I contend that the possibility is opened in which we can 'free thought from what it silently thinks, and so enable it to think differently' (Foucault 1992: 9). To free thought and think differently about health, bodies and life, it is necessary to expose and critique the rationalities that seek to control and regulate individuals and populations.

To achieve this I employ Foucault's analysis of biopolitics to examine 'the set of mechanisms through which the basic biological features of the human species became the object of a political strategy' (Foucault 2007: 1). Specifically, I investigate the set of lifestyle mechanisms within which everyday and biological life of the individual and population become the object of political strategy. Further, I develop Foucault's notion of *dispositif* as an analytic framework for tracing the emergence of lifestyle. *Dispositif*, or what I call an 'enabling network', is an entanglement of the three main lines of Foucault's work: knowledge, power and subjectivity. The enabling network operates in response to 'an *urgent need*' which activates or enables its 'dominant strategic function' (Foucault 1980: 195). In response to the urgent need of securing the threat of obesity, the network of lifestyle enables health campaigns, experts, medical professionals, magazines, tape measures, smartphones and health policies as biopolitical strategies. Without the urgency of obesity or the entangled epidemiological and economic knowledges, these strategies would not be enabled or activated but remain dormant. The urgent need varies between historical periods. It could be mental illness, smallpox, the masturbating child, refugees or, in this case, obesity. It is around the urgency of obesity that the imperative of the lifestyle network to govern the choices and bodies of individuals is created.

The objective of this book is not to argue that governments should refrain from interfering with individual choices or lives (this would be the aim of a healthism critique). My objective is to demonstrate the way lifestyle is used as a biopolitical mechanism. Examined through a biopolitical lens, lifestyle can be seen as a strategy that calls into question and makes visible the everyday lives of some individuals. If the individual *appears* to make healthy choices, they remain inconspicuous and are enfolded into the secured population. However, if the individual *appears* to choose irresponsibly, particularly by having an 'overweight' body, then their life is made conspicuous as a potential or actual threat to population health. By examining lifestyle through this lens, I attempt to address recent governmental trends that use 'choice' to call into question some lives and justify narrowing the circle of collective welfare in a manner that excludes those that are perceived to choose irresponsibly.

Considering the diverse literature and topics analysed in this book it is important to be clear about the approach I adopt. This book does not sit comfortably within particular disciplinary boundaries. Although I primarily draw on the philosophy of Foucault, I also appeal to sociology, public health, economics, political science and feminist literatures to provide a unique and distinct analysis of lifestyle and its role in society. In drawing on these disciplines, I argue that lifestyle is composed of 'lines of visibility and enunciation, lines of force, lines of subjectification, lines of splitting, breakage, fracture, all of which criss-cross and mingle together' (Deleuze 1992: 162). The 'crisscrossing', 'fracturing' and 'intermingling' lines contributed by different disciplines, discourses, objects and tools of the network of lifestyle necessitate an interdisciplinary mode of analysis. This approach charts the constitution of lifestyle as a multi-linear network of relations, practices and knowledges through which everyday lives, choices and bodies are made visible and governable. In adopting an interdisciplinary analytic it is possible to tease out and trace the 'crisscrossing' lines comprising the lifestyle network. This approach makes it possible to investigate the relationship between lifestyle and problems of individual freedom, governmentality and subjectivity in neoliberal societies. Furthermore, a critical investigation into the biopolitical use of the lifestyle network opens the possibility for norms and relations to be questioned, with an opportunity for producing new norms of life.

This book is divided into seven chapters. Chapter 1 develops the theoretical and philosophical tools necessary for examining lifestyle as a biopolitical mechanism. I open this chapter by focusing on bioethical arguments against obesity. Prominent bioethicists Peter Singer and Dan Callahan have both argued that obese people burden and harm society and therefore governments are justified to coerce or even stigmatize people who are overweight or obese. I use these arguments as examples to highlight the way obesity is understood as an individual problem and the perceived threat of individual choices and bodies to society. Following this example, I interpret and develop Foucault's notion of *dispositif* as an enabling network through which individual choice and behaviour is made visible and governable. I draw together *dispositif* with his analysis of biopolitics and show why a biopolitical analysis is able to provide a unique perspective on the governmental use of lifestyle.

Chapter 2 emphasizes the role of individual choice and responsibility in neoliberal theory and the way the lifestyle network problematizes choice as a security issue requiring biopolitical governance. I identify three tensions in the biopolitical governance of choice: present and future, freedom and security, individual and population. I then demonstrate how 'healthy subjects' are produced and enfolded into the secured population, whereas 'irresponsible subjects' are excluded. I conclude the chapter with an analysis of techniques of governance employed in the Australian government's *Measure Up* social marketing campaign. I use this campaign as an illustration of neoliberal health policies that seek to govern individual choice and bodies toward norms of health as a means of securing the population.

Chapter 3 sketches the intermingling of neoliberal ideas of governance with epidemiology and health policy that make the daily practices and choices of the individual visible as biopolitical targets. In locating the cause of epidemics in daily practices and choices of the individual, lifestyle epidemiology makes activities, bodies and behaviours visible for biopolitical mechanisms to target and govern. The objective of this chapter is to trace the thread of epidemiology and health promotion through the fabric of the lifestyle network.

Chapter 4 introduces the aesthetics of the body and choices. The visibility of the body and choices as indicators of individual beauty and health as well as responsibility to the population enables lifestyle to govern individuals through consumption and identity creation. The sociological work of Henri Lefebvre, Guy Debord and Pierre Bourdieu is critically incorporated with Foucault's ethics of the self to analyse the aesthetic thread in the lifestyle network. The purpose of this chapter is to bring sociological and governmental understandings of lifestyle into closer conversation to highlight the way individuals are enticed towards particular subject positions.

Chapter 5 examines the abundance of lifestyle guidance available through print and social media, and the way this dovetails with traditional sources of health guidance, such as doctors and health departments. Drawing on examples from Jamie Oliver, Michelle Obama, smartphone apps, and *Prevention* magazine, I argue that the urgency surrounding obesity serves as a focal point that harmonizes the cacophonous voices of lifestyle guidance into a governmental network. These knowledges and practices are not forced on the individual but purchased and consumed in a bid to create a healthy and beautiful lifestyle. This chapter demonstrates the way medical and non-medical experts function as biopolitical pastors who hear confessions, circulate knowledge and deploy techniques to shape subjects in the lifestyle network.

Chapter 6 explores the possibility of critique and resistance within the lifestyle network. The lifestyle network enables the shaping and styling of everyday practices into subjectivities that adhere to biological and social norms. The successful adoption of these norms makes the individual visible as a healthy subject. However, the norms and strategies deployed throughout this network are not neutral or beneficial for all individuals. Failure to adhere to these norms can result in disciplinary mechanism and withdrawal of protection and care. To illuminate this discussion I draw on the Health at Every Size and fat acceptance movements as examples of practical strategies of resistance.

Chapter 7 extends the analysis from Chapter 6 to argue that resistance needs to come from individuals in relations of care with others, which can establish new modes of living together. Contrary to popular criticisms of Foucault, this chapter demonstrates that the practices of the self are not individualistic but are deeply inter-subjective and can provide important clues for collective resistance to the biopolitical mechanisms of governance that focus on individual choices, behaviours and bodies. I explore the similarities between pastoral relations and care of the self relations to emphasize the importance of masters or mentors in resistance. Finally, I argue that this form of resistance will be at the level of the everyday, restless and communal.

Before closing this Introduction I wish to offer a brief statement on what this book is and what it is not. This book is not anti-health or anti-medicine. Nor is it anti-government or anti-authority. My purpose here is subtler. I seek to provide a critical analysis of the norms of bodies and health circulated through the lifestyle network that are used to govern individuals in the name of population and economic security. Through this critique I hope to open spaces for different norms of life that enable individuals to relate with others in styles and ways that allow a diversity of bodies and experiences of health that are not governed so much by a plurality of means toward a limited number of alien and abstract ends.

## References

Abercrombie, Nicholas, Stephen Hill, and Bryan S. Turner, eds. 2006. *The Penguin Dictionary of Sociology*. 5th edn. Harmondsworth: Penguin.

Baum, Fran. 2011. "From Norm to Eric: avoiding lifestyle drift in Australian health policy." *Australian and New Zealand Journal of Public Health* 35 (5):404–406. doi: 10.1111/j.1753-6405.2011.00756.x.

Bealey, Frank. 1999. *Blackwell Dictionary of Political Science*. Oxford: John Wiley & Sons.

Brownell, Kelly D., Rogan Kersh, David S. Ludwig, Robert C. Post, Rebecca M. Puhl, Marlene B. Schwartz, and Walter C. Willett. 2010. "Personal responsibility and obesity: a constructive approach to a controversial issue." *Health Affairs* 29 (3):379–387. doi: 10.1377/hlthaff.2009.0739.

Callahan, Daniel. 2013. "Obesity: chasing an elusive epidemic." *Hastings Center Report* 43 (1):34–40.

Campos, Paul. 2004. *The Obesity Myth: Why America's Obsession with Weight Is Hazardous to Your Health*. New York: Gotham Books.

Carter, Stacy M., Lucie Rychetnik, Beverley Lloyd, Ian H. Kerridge, Louise Baur, Adrian Bauman, Claire Hooker, and Avigdor Zask. 2011. "Evidence, ethics, and values: a framework for health promotion." *American Journal of Public Health* 101 (3):465–472. doi: 10.2105/ajph.2010.195545.

Crawford, Robert. 1980. "Healthism and the medicalization of everyday life." *International Journal of Health Services : Planning, Administration, Evaluation* 10 (3):365–388.

Deleuze, Gilles. 1992. "What is a dispositif?" In *Michel Foucault, Philosopher*, edited by Timothy J. Armstrong. New York: Routledge.

Department of Health. 2011a. *Healthy Lives, Healthy People: A Call to Action on Obesity in England.* edited by Department of Health. London: Crown.

Department of Health. 2011. *The Public Health Responsibility Deal*. Crown 2011 [cited February 1 2012]. Available from http://www.dh.gov.uk/prod_consum_dh/groups/dh_digitalassets/documents/digitalasset/dh_125237.pdf.

Fitzgerald, Faith T. 1994. "The tyranny of health." *New England Journal of Medicine* 331 (3):196–198. doi: doi:10.1056/NEJM199407213310312.

Fitzpatrick, Michael. 2001. *The Tyranny of Health: Doctors and the Regulation of Lifestyle*. London: Routledge.

Flegal, Katherine M., Brian K. Kit, Heather Orpana, and Barry I. Graubard. 2013. "Association of all-cause mortality with overweight and obesity using standard body mass index categories: A systematic review and meta-analysis." *JAMA* 309 (1):71–82. doi: 10.1001/jama.2012.113905.

Foucault, Michel. 1980. "The confession of the flesh." In *Power/Knowledge: Selected Interviews and Other Writings*, edited by Colin Gordon. New York: Pantheon Books.

Foucault, Michel. 1992. *The Use of Pleasure: The History of Sexuality Volume 2*. Translated by Robert Hurley. Harmondsworth: Penguin Books.

Foucault, Michel. 2007. *Security, Territory, Population: Lectures at the Collège de France 1977–78*. Translated by Graham Burchell. Edited by Arnold I. Davidson. New York: Palgrave Macmillan.

Gard, Michael. 2011. *The End of the Obesity Epidemic*. New York: Routledge.

Gordon-Larsen, Penny, Melissa C. Nelson, Phil Page, and Barry M. Popkin. 2006. "Inequality in the built environment underlies key health disparities in physical activity and obesity." *Pediatrics* 117 (2):417–424. doi: 10.1542/peds.2005-0058.

Hagan, Kate. 2012. "Call for school weigh-ins to fight fat." *Sydney Morning Herald*, August 27, 2012.

Howell, Jeremy, and Alan Ingham. 2001. "From social problem to personal issue: the language of lifestyle." *Cultural Studies* 15 (2):326–351.

Illich, Ivan. 1975. *Medical Nemesis: The Expropriation of Health*. London: Calder & Boyars.

Lacy, K., P. Kremer, A. de Silva-Sanigorski, S. Allender, E. Leslie, L. Jones, S. Fornaro, and B. Swinburn. 2012. "The appropriateness of opt-out consent for monitoring childhood obesity in Australia." *Pediatric Obesity* 7 (5) doi: 10.1111/j.2047-6310.2012.00076.x.

Lang, Tim, and Geof Rayner. 2010. "Corporate responsibility in public health." *BMJ* 341. doi: 10.1136/bmj.c3758.

Leonhardt, David. 2009. "The way we live now – fat tax." *New York Times*, August 12.

Lowe, Adrian. 2012. "Is this child abuse? The courts think so." *The Age*, July 12.

Marmot, Michael. 2000. "Social determinants of health: from observation to policy." *Medical Journal of Australia* 172 (8):379–382.

Murray, Samantha. 2008. *The 'fat' female body*. Basingstoke: Palgrave Macmillan.

Murtagh, Lindsey, and David S. Ludwig. 2011. "State intervention in life-threatening childhood obesity." *JAMA* 306 (2):206–207. doi: 10.1001/jama.2011.903.

National Preventative Health Taskforce. 2009. *Obesity in Australia: A Need for Urgent Action.* edited by Department of Health and Ageing. Canberra: Australian Government.

Office of the First Lady. 2010. *Remarks of First Lady Michelle Obama – Let's Move Launch*. The White House 2010 [cited January 18 2012]. Available from http://www.whitehouse.gov/the-press-office/remarks-first-lady-michelle-obama.

Porta, Miquel, ed. 2008. *Dictionary of Epidemiology*. New York: Oxford University Press.

Roberts, Ian, and Phil Edwards. 2010. *The Energy Glut: Climate Change and the Politics of Fatness*. London: Zed Books.

Singer, Peter. 2012. Weigh more, pay more. *Project Syndicate: A World of Ideas*, http://www.project-syndicate.org/commentary/weigh-more-pay-more.

Skrabanek, Petr. 1994. *The Death of Humane Medicine and the Rise of Coercive Healthism*. London: Social Affairs Unit.

Wann, Marilyn. 2009. "Foreword – fat studies: an invitation to revolution." In *The Fat Studies Reader*, edited by Esther Rothblum and Sondra Solovay. New York: New York University Press.

Wen, Li Ming, Louise A. Baur, Judy M. Simpson, Chris Rissel, and Victoria M. Flood. 2011. "Effectiveness of an early intervention on infant feeding practices and 'tummy time': a randomized controlled trial." *Archives of Pediatric and Adolescent Medicine* 165 (8):701–707. doi: 10.1001/archpediatrics.2011.115.

World Health Organization. 2000. *Obesity: Preventing and Managing the Global Epidemic*. World Health Organization technical report series. Geneva: WHO

# 1 Obesity, bioethics and the lifestyle *dispositif*

Personal responsibility is the foundation of an ethical society.
David Cameron, UK Prime Minister (2010).

In its everyday usage, the concept of lifestyle appears to refer to a set of stable and concrete phenomena. Politicians valorize their nation's lifestyle as uniquely good and worthy of protection, fashion designers claim their clothes and accessories provide a lifestyle, medical professionals fear the effect of the modern lifestyle on global health, and environmentalists chide the Western lifestyle as a primary cause of climate change. In these instances the intended audience is assumed to know what is being referred to, however, closer inspection reveals lifestyle as an imprecise and vague idea.

Lifestyle may have an ambiguous meaning, but like life, health or sex, it has a powerful rhetorical influence. Lifestyle is increasingly used to frame political action, shape social relations, and redefine interactions between individuals and populations. The diverse uses of lifestyle establish a network through which the everyday habits and activities of individuals are made visible and governable. Food choice, exercise habits, fashion or leisure activities *may* be innocuous personal preferences, but viewed through the lens of lifestyle they represent an identity, a politics, or a threat to population health. In making these choices visible, techniques of governance can be mobilized to secure the population by encouraging individuals to adopt responsible lifestyle choices. The successful creation of a healthy lifestyle enfolds the individual into the population to be protected and cared for. However, failure to adopt choices associated with a healthy lifestyle can result in the stigmatization and scrutiny of the individual's life through disciplinary mechanisms.

The biopolitics of lifestyle masks a shift from a socialized welfare that purports to secure the many, to an individualized welfare that secures those who are able to care for themselves. Lifestyle makes visible those who are responsible and those who purportedly harm society. To be seen as harming or financially burdening other can lead to implicit or explicit exclusion from care. Stigma and harm associated with lifestyle is an underexplored theme. Much of the discussion of the purported harmful effects of lifestyle is taken for granted. It is assumed as an obvious fact that smokers, the obese and the lazy are a drain on society. This allows the effects of stigma to be ignored or minimized.

This chapter uses a recent debate in the bioethics literature over obesity and the individual to illustrate the way the logics of lifestyle allows certain interventions to prevent harms seem obvious and acceptable. The obviousness of these interventions hides and minimizes the stigmatization of people who are overweight or obese. Prominent bioethicists Peter Singer and Dan Callahan have argued for coercive interventions into the lives of individuals on grounds that unhealthy choices of individuals harm the rest of the population. This bioethical debate serves to highlight the usefulness of the work of Michel Foucault in tracing the entangled lines and threads that flow through strategies of lifestyle governance.

Following an analysis and discussion of Singer and Callahan's arguments, I develop Foucault's notions of *dispositif* and biopolitics. *Dispositif* is an apposite analytic framework for this project as it incorporates the three main lines of Foucault's work – knowledge, power, and subjectivity – to make specific subjects, truths, and practices visible, possible, and governable. However, it is under-theorized in existing literature and mistakenly assumed to be equivalent to apparatus or indistinct from genealogy. Although related to both apparatus and genealogy, my analysis sheds light on the concept of *dispositif* in Foucault's later work. I use this broadened perspective to address the way lifestyle is a biopolitical mechanism that identifies and governs those whose lives are to be cared for and those excluded from care and exposed to violence. This chapter sets up the theoretical groundwork to respond to the questions of how lifestyle emerged to govern, guide and foster the life of the individual for the security of the population.

## Bioethics and the harm of obesity

Debate over the impact of the obesity epidemic and what should be done about it has been ongoing among public health, health policy and social scientists since at least the 1990s. Bioethicists, however, have been relatively silent on this issue. Traditionally focused on ethical issues pertaining to the individual and the clinic, bioethicists have addressed obesity primarily in the context of concerns relating to resource allocation or clinical procedures such as bariatric surgery. Prominent bioethicists Peter Singer and Dan Callahan have recently entered the obesity debate to argue that obesity is not simply a clinical or personal issue but an ethical issue with social and political consequences (Singer 2012, Callahan 2013). As such, they contend that interventions into the lives of individuals and populations are needed to curb the rise of obesity.

The arguments put forward by Singer and Callahan are not particularly striking or original. What is interesting however is the use of normative ethics to justify coercive interventions as well as the reliance on the logics of lifestyle for these arguments and interventions to seem like common sense. The strength of their respective arguments rests on empirical claims that obesity is caused by individual liberty, which in turn harms society. Although tightly argued, Singer and

Callahan's proposals do not accord with empirical research. By drawing attention to this research I argue that Singer and Callahan contribute to the stigmatization of people defined as overweight or obese, which can itself be considered a form of harm.

The contributions of Singer and Callahan to the obesity debate provide a helpful avenue into the way obesity is framed such that individuals' incidental behaviours become visible while more problematic structural forces remain hidden. Singer and Callahan's respective arguments that individuals should be made responsible for the obesity epidemic contain two problems widespread in obesity discourse. First, an uncritical assumption that individuals are autonomous agents responsible for health related effects associated with food choices and physical inactivity. In their view, individuals are obese because they choose certain foods or refrain from physical activity. However, they acknowledge that this alone does not justify intervening in the lives of individuals. Both Singer and Callahan believe that autonomous individuals are free to make foolish choices so long as those choices do not harm or impose costs on others. It is at this point that the second problematic aspect arises. To legitimately interfere in the liberty of individuals they invoke the harm principle. That is, they seek to demonstrate that obese individuals are harming others. In attempting to activate the harm principle, both Singer and Callahan rely on superficial readings of public health research to amplify the harm caused by obese individuals and ignore pertinent epidemiological research on the social determinants of obesity. The mobilization of the harm principle and corresponding focus on individual behaviours is made possible through the belief that the individual is the primary unit through which to understand social interactions.

Like many within contemporary bioethics, Singer and Callahan operate with a theoretical toolbox largely defined by the liberal tradition of moral and political philosophy. In particular, they characterize persons as rational, self-conscious and autonomous agents free to determine their own ends. This idea is upheld as axiomatic and deserving of social and juridical guarantee. The dominance of this idea among bioethicists partly explains their relative silence on public health issues like obesity. If it is true that obesity is the result of choosing to eat more and exercise less, then it would appear that autonomous persons in liberal societies should be left alone to pursue the activities and ends of their own determining. This at least seems to be the conclusion to draw from J.S. Mill, who writes 'the only purpose for which power can be rightfully exercised over any member of a civilized community, against his [*sic*] will, is to prevent harm to others. His own good, either physical or moral, is not a sufficient warrant' (1996: 78). However, as is demonstrated below, the relationship between individuals and society is not so clear-cut.

### *Singer on the cost of heavy passengers*

In an article for *Project Syndicate*[1] entitled 'Weigh More, Pay More', Singer echoes the harm principle when he writes:

Is a person's weight his or her own business? Should we simply become more accepting of diverse body shapes? I don't think so. Obesity is an ethical issue, because an increase in weight by some imposes costs on others.

(Singer 2012)

Singer uses the example of air travel to demonstrate the way obese individuals impose a cost on others. Drawing on figures offered by the chief economist at Qantas, Singer claims that a weight increase of two kilos per passenger since 2000 has resulted in the airline spending an extra $472 in fuel per Sydney to London flight or $1 million annually. Singer suggests that this cost should not be borne by the airline or the passengers as a collective, but by those individuals exceeding a 75-kilo threshold. Singer claims that this measure is 'not to punish a sin' but 'a way of recouping…the true cost of flying you to your destination, rather than imposing it on your fellow passengers' (2012).

Singer points out that air travel is not a human right and that appropriate costs for air travel is limited to the realm of private enterprise. If Qantas wishes to pass operating costs on to customers it could. To this point Singer is sounding more like a spokesperson for the airline industry rather than a bioethicist. However, Singer's real concern is not Qantas's bottom line. Although the airplane example and title of his article suggests that Singer wants obese people to pay the financial costs purportedly associated with their weight, his real interest is in non-financial harms.

The more obviously ethical feature of Singer's argument is when he links increased fuel use to global warming. According to Singer, heavy passengers require more fuel, which produces 'higher greenhouse-gas emissions' and 'exacerbate[s] global warming' (2012). He applies the same logic to public transport and health care. According to Singer, larger and heavier people use more health care resources due to their greater mass and the variety of medical problems purported to be associated with being overweight and obese.

Singer maintains that these examples demonstrate that 'the size of our fellow-citizens affects us all' and that '[i]f we value both sustainable human well-being and our planet's natural environment, my weight – and yours – is everyone's business' (2012). Singer concludes by returning to the harm principle, arguing that the harms and costs caused by obese individuals 'justify public policies that discourage weight gain' (2012). Thus it is not simply a matter of requiring obese individuals to offset the economic costs, but due to non-economic costs such as harm to the environment or use of health resources, it is necessary to discourage and prevent individuals from making choices purported to cause obesity. It is on this point that Callahan offers a number of suggestions, including stigmatization.

### Callahan and the use of stigma to benefit society

In an article published in the *Hastings Center Report*, Daniel Callahan asks 'how far can government and business go in trying to change behavior that harms health?' (2013). Callahan is specifically interested in behaviours associated with obesity. Like Singer, Callahan considers obesity to pose significant harms to human wellbeing and society. According to Callahan, the combination of prevalence (67 per cent of Americans are overweight or obese) and costs (financial, social and medical) of obesity constitutes a harm that justifies the introduction of 'coercive public health measures' (2013: 36).

The nature of the problem, as defined by Callahan, means that individuals need to be induced 'to change the way they eat, work, and exercise' (2013: 39). Despite acknowledging social influences, Callahan argues that the primary 'causes of obesity' are the sedentary habits, poor diet and 'all the luxuries we possess – automatic garage door openers, can openers, food blenders and mixers, escalators, elevators, golf carts, automobiles, and so on' (2013: 35). Therefore it is at this level that behaviour change strategies need to operate and 'awaken' overweight and obese people 'to the reality of their condition' (2013: 40).

What is needed, according to Callahan, is a three-pronged approach that includes 'coercive public health measures, mainly by government but also by the business community; childhood prevention programs; and social pressure on the overweight' (2013: 36). Although Callahan notes a role for government and business in shaping the environment, he maintains that ultimately individuals need to work 'to stay thin in the first place and to lose weight early on if excess weight begins to emerge' (2013: 40). By coercing and pressuring individuals, Callahan believes it will be possible to change behaviours and reduce obesity rates. Callahan acknowledges that these strategies will interfere with individual liberty, however, like Singer, Callahan believes that the magnitude of the harms posed to future human wellbeing and society justify this approach.

Both Singer and Callahan assert that obesity harms society due to associated economic costs. A strategy to rebut their respective arguments could be to refute this economic assertion. The *Wall Street Journal* published an estimate that the diet and obesity management industry will be worth $139.5 billion dollars by 2017 (Anonymous 2013). This figure does not include the value of so-called obesogenic industries such as fast food and soda. It could therefore be argued that the health costs of obese people and associated activities are neutralized by the gains of industries associated with causing or curing obesity. Rather than targeting individuals, Singer and Callahan should focus their energies on getting governments to adjust tax rates and distribution accordingly to ensure public transport and health systems can grow to support these profitable industries. Rather than taking this route I will assume that the collective weight gain in Western populations poses some economic costs to society. However, I maintain that the positions of Singer and Callahan are untenable and ignore pertinent research. To rebut their positions I focus on two related questions: Do individual choices cause obesity? And, are the strategies targeting the individual ethically justified?

The question of individual choice has become a vexing issue in public health in Western societies. Daniel Goldberg argues that a 'methodological individualism' dominates much of debate around public health interventions, particularly in the United States. Drawing on Jon Elster (1982), Goldberg defines methodological individualism as 'the doctrine that all social phenomena (their structure and their change) are in principle explicable only in terms of individuals – their properties, goals, and beliefs' (Goldberg 2012: 104). Methodological individualism is a key doctrine in neoliberal governmentality, to be discussed in the following chapter. This principled or philosophical position has led to lifestyle-focused health promotion strategies and policies that focus almost exclusively on individual behaviours. The focus on the individual in matters of public health however, owes more to the philosophical and political significance of the individual in the West and neoliberal theories than empirical evidence. An increasing body of research demonstrates that health in general, and obesity in particular, is the 'result of the way in which we organize our societies through economic, social, and political policies and practices' (Venkatapuram, Bell, and Marmot 2010). While Singer and Callahan emphasize individual choice as the cause of obesity, and therefore a legitimate target for interventions, public health research demonstrates that social and structural factors preclude and shape the possibility of choice.

Over the past decade public health associations and research institutions have emphasized that social, cultural, economic and environmental factors are antecedent to individual behaviours associated with obesity (Rudd Center 2013, Public Health Association of Australia 2010). Leaders in the field, such as Michael Marmot in the UK, Fran Baum in Australia, and Kelly Brownell in the US, all argue that structural factors rather than individual choice should be the main target of interventions to prevent obesity (Marmot 2010, Baum and Sanders 2011). Brownell and his colleagues write:

> environmental conditions can override individual physical and psychological regulatory systems that might otherwise stand in the way of weight gain and obesity, hence undermining personal responsibility, narrowing choices, and eroding personal freedoms.
>
> (Brownell et al. 2010)

Individual behaviours do have a part to play in health status, but focusing on the individual in the absence of the social determinants of obesity will not address the complexity of the issue. Marmot contends that 'policies to modify health behaviors need to address the social determinants of health. Aiming interventions at individuals will not by themselves reduce health inequalities' (2010: 141). For Marmot, any focus on the individual needs to occur within a broader strategy that addresses social determinants. However, the influence of methodological individualism and the logics of lifestyle make such approaches unlikely to eventuate. As will be discussed throughout this book, recent obesity interventions in Australia, the UK and the US have almost exclusively focused on individual consumer choices with no substantial effort made by governments to address social determinants (Baum 2011).

Not only is focusing on individual choice as the cause of obesity unsupported by research, but research also demonstrates that choice is undermined by the social factors that many consider to be the conditions of the obesity epidemic. Writing in the *New England Journal of Medicine*, Jennifer Cheng describes her experience as a pediatrician filing a Child Protective Services (CPS) report for medical neglect of two young girls with progressive morbid obesity. Cheng soon came to realize however, that interventions such as CPS and strategies focused at the choices and behaviours are ineffective if those interventions do not address structural factors of poverty, crime and education. Cheng writes:

> People born into lower social strata are more likely than their contemporaries in higher social echelons to be born small and then to experience rapid catch-up growth leading to overweight and obesity; they also have higher rates of pulmonary and cardiovascular disease, learning difficulties, mental illness, poor life quality, and premature death than do people higher up the social ladder... Making the right decisions can be extraordinarily difficult for families like the S. family, because they have little true choice.
>
> (Cheng 2012)

The research of Marmot, Brownell, Baum and Cheng represents a much wider body of research that repeatedly demonstrates that individual choices cannot be held as the exclusive cause of an individual's obesity and associated health status (Galer-Unti 2013). Furthermore, many of the choices available to individuals are shaped and influenced by social, political and economic factors beyond the individual's control. Even if obesity does harm society as Singer and Callahan contend, the public health research is clear that a plurality of factors operating prior to individual choice is responsible for that harm.

In light of this public health research, the second question – Are the strategies targeting the individual ethically justified? – must be answered in the negative. If individual liberty is not the primary cause for the obesity-related harms to society, and that focusing on the individual will not reduce those harms, then the strategies targeting the individual outlined by Singer and Callahan do not fulfill the criteria of the harm principle. However, the importance of this conclusion is not that Singer and Callahan are simply wrong on etiological grounds. The more troubling point is that the kinds of arguments put forward by Singer and Callahan contribute to the harms suffered by individuals. Yet due to the biopolitics of lifestyle some activities, behaviours or bodies are made visible while others are hidden. The societal harms purported to be associated with obesity serve to mobilize a range of governmental strategies, while the exclusion and stigmatization that occurs through these processes is considered to be a further consequence of the initial irresponsibility of the individual.

In attempting to invoke the harm principle to justify interventions into the lives of individuals, Singer implies and Callahan outlines, strategies that are known to cause serious harm. Not only are attempts to hold individuals responsible for body weight unsupported by evidence and demonstrated to be ineffective, but such interventions contribute to stigmatization (Goldberg and Puhl 2013, Guttman and Salmon 2004). According to Goldberg and Puhl, weight stigmatization has 'numerous adverse health consequences, including depression, anxiety, low self-esteem, suicidal ideation, and avoidance of health care' (2013: 6). Lily O'Hara and Jane Gregg also demonstrate that body dissatisfaction, disordered eating, discrimination and death are some of the iatrogenic effects of health promotion strategies targeting individual behaviours in order to reduce body weight (O'Hara and Gregg 2006: 261).

If the public health research cited here demonstrates anything, it is that obesity as an individual condition and the obesity epidemic as a global phenomenon is extremely complex. The editors of the *New England Journal of Medicine* cautioned in 1998, at the start of the so-called obesity epidemic, that 'Until we have better data about the risks of being overweight and the benefits and risks of trying to lose weight, we should remember that the cure for obesity may be worse than the condition' (Kassirer and Angell 1998). Fifteen years later, a systemic review by Katherine Flegal and co-workers published in the *Journal of the American Medical Association* suggests that being overweight or slightly obese is negligible and in some cases beneficial (Flegal et al. 2013). In contrast, the so-called cures discussed here are well documented to cause harm.

These 'ethical' arguments are framed and enabled by the rationale of lifestyle. However, to fully grasp and critique the biopolitical nature of these bioethical arguments and public health policies we need to move beyond the tools and concepts of the moral and political philosophy of the liberal tradition with its overemphasis on individual autonomy (Mills 2010, Dawson 2010, Bishop and Jotterand 2006). It is here that Michel Foucault's concept of the *dispositif* and analyses of biopolitics become crucial.

## *Dispositif* as an enabling network

In his essay 'What is a *dispositif*?' Gilles Deleuze describes Foucault's concept as a 'tangle' and 'multilinear ensemble' of different lines that produce a heterogeneous network of objects, subjects, discourses and practices (Deleuze 1992: 159). According to Deleuze, the *dispositif* can be thought of as 'machines which make one see and speak' (knowledge), 'lines of force...[that] proceed from one unique point to another' (power), and 'lines of subjectification' that allow subjectivity 'to come into being or makes it possible' (1992: 161). Deleuze's articulation of *dispositif* as incorporating the major aspects of Foucault's work – knowledge, power, and subjectivity – resonates with Foucault's own articulation.

The late 1970s was a period of transition for Foucault. Broadening his notion of '*episteme*' – developed in *The Order of Things* – the concept of *dispositif* framed Foucault's analyses of discipline, sexuality and biopolitics (Veyne 2010: 93). While episteme focused on discourse, rationality and the scientific production of true and false categories of knowledge, *dispositif* is 'both discursive and non-discursive' (Foucault 1980: 197). The disciplinary *dispositif* of the penal system employs a matrix of non-discursive practices such as the prison architecture, timetable and most famously the panopticon. Dreyfus and Rabinow describe the *dispositif* as bringing 'together power and knowledge into a specific grid of analysis' (1983: 121). Foucault's most detailed account of the *dispositif* is in the conversation 'The Confession of the Flesh' (1980). Here

Foucault outlines three features. The first feature is a 'heterogeneous ensemble consisting of discourses, institutions, architectural forms, regulatory decisions, laws, administrative measures, scientific statements...in short, the said as much as the unsaid' (1980: 194). The second feature is the agonistic logic that holds heterogeneous elements, knowledges and practices together. The final feature is 'responding to an *urgent need*' (Foucault 1980: 195). While the first two features have been discussed extensively in secondary literature, responding to an 'urgent need' is often ignored or assumed. The 'urgent need' is historically contingent and serves as an imperative that mobilizes the strategic networks of power, knowledge and subjectification. The idea of an urgent need is important on two fronts: i) in the response, which suggests the active and provocative force of life; and ii) the urgent need activates the *dispositif,* without which it would be a dormant system. I expand on these two points shortly.

Unfortunately there is not a satisfactory English translation for *dispositif.* This gives rise to the temptation to use the French. However, to use *dispositif* exclusively runs the risk of alienating readers, or worse, transforming a flexible analytic tool into a rigid abstraction. The term 'dispositive' is the closest literal English translation, however there is debate as to whether this adequately grasps the variety of Foucault's usage.[2] *Dispositif* has been translated as 'deployment' by Robert Hurley in *The Will to Knowledge* (Foucault 1998: 106), or as 'apparatus' in the Graham Burchell translations of the *Collège de France* lectures (Foucault 2006, 2005, 2007). In translating Gilles Deleuze's essay, 'What is a *dispositif*?', Timothy J. Armstrong provides a helpful note, stating, '[t]here is, in English, no straightforward way of translating *dispositif*...I have used the terms "social apparatus" or "apparatus" as the closest available equivalent' (Deleuze 1992: 159). Apparatus, however, is not an adequate solution.

Despite conceptual difficulties, 'apparatus' is increasingly gaining a foothold as the preferred term for *dispositif.* Jeffrey Bussolini notes that apparatus has received acceptance partly owing to the influence of Giorgio Agamben's *What is an Apparatus?* (2009).[3] However, an initial difficulty with rendering *dispositif* as apparatus, acknowledged by Burchell, is that apparatus is already used to translate *appareil* (Foucault 2006: xxiii). To use apparatus to translate *dispositif* 'risks confusion with, for example, "State apparatuses" (*appareils d'Etat*), from which Foucault clearly wants to distinguish his own analysis' (Foucault 2006: xxiii). While apparatus has a certain resonance in English, rendering both *dispositif* and *appareil* as apparatus 'collapses distinct conceptual lineages...and produces a false identity' (Bussolini 2010: 106). Rather than sharing an identity, *dispositif* and *appareil* are used by Foucault in different, yet related ways. The relationship between *dispositif* and *appareil* is made clearer when their respective meanings and designations are understood. The concept *appareil* is clearly entwined with *dispositif,* and plays an important role in defining its function. However, it is also clear that Foucault does not regard or use the concepts synonymously. *Appareil* is a separate yet linked subset of *dispositif.* Burchell suggests that Foucault uses *dispositif* to 'designate a configuration or arrangement of elements and forces, practices and discourses, power and knowledge, that is both *strategic* and *technical*' (Foucault 2006: xxiii). *Appareil*, however, is more restricted, insofar as it is defined by and affiliated with a *dispositif* or institution such as the State. Both *appareil* and *dispositif* relate to 'machine' or 'device'. However, Bussolini's analysis of the French etymology defines *appareil* as a particular thing – a telephone, aircraft or camera – whereas *dispositif* relates to 'the enacting terms (of a law or decision), disposition of troops in battle, or a device or contrivance' (Bussolini 2010: 95). *Dispositif* conditions the activity of the device or apparatus as it interacts with other elements. In this way *dispositif* is closer to Hurley's translation as deployment. While Foucault is not always consistent in his own usage it is clear that there are significant conceptual reasons for distinguishing the terms.

Wishing to avoid reliance on the French or resignation to an infelicitous English translation, Dreyfus and Rabinow suggest 'grid of intelligibility'. However, this rendering could be read as overemphasising knowledge. As they acknowledge themselves, it does not account for what Foucault was trying to reveal about practices (Dreyfus and Rabinow 1983: 121). The importance of practices is particularly relevant in relation to the 'urgent need' that the *dispositif* responds to. I suggest that instead of grid of intelligibility, the *dispositif* can be conceived as an 'enabling network'. This network of power, knowledge, and subjectivity enables certain practices, bodies and lives to be visible and governable. Although not a direct translation, enabling network captures the conceptual emphasis on practices and the reactiveness of biopolitics. This becomes clear through a discussion of two aspects of the 'urgent need' of the *dispositif*: responsiveness and activating.

### Responsiveness

In 'The Confession of the Flesh' Foucault uses the example of madness to illustrate his point about the 'urgent need'. The urgent need varies between historical periods. It could be smallpox, refugees, the radicalization of Muslim youth or body weight. However, a more general example that I draw on is Foucault's discussions of life as exceeding the grasp of biopolitics. Life has emerged as an urgent need that provokes the response of biopolitical strategies of control. According to Catherine Mills, this is a point that has been neglected due to a lack of attention to the norms of biological life and their relation to the norms of social life. I briefly address the role of norms of life to expand on the importance of the 'urgent need' in the *dispositif* as an enabling network.

In the enthusiasm to embrace the idea of biopolitics, much of the secondary literature focuses on the politics of biopolitics, while assuming what the 'bio' or 'life' is and does. However, it is through understanding the 'life' or bio of biopolitics that the political strategies of governance and control can be understood. Mills argues that understanding the way life and politics combine to form 'biopolitics' enables the recognition of the 'active power of life itself' (2013: 90). An important influence on Foucault's thinking about life and norms is Georges Canguilhem. Foucault discusses Canguilhem in a number of texts, most relevant here are his comments on normalization in *Abnormal: Lectures at the*

*Collège de France 1974–1975* and in a 1983 essay on life used as an introduction to the 1989 English translation of Canguilhem's *The Normal and the Pathological*. In discussing Canguilhem and the norms of life in *Abnormal*, Foucault states that the norms function 'is always linked to a positive technique of intervention and transformation, to a sort of normative project' (Foucault 2000b: 50). According to Mills, 'Foucault's account of the operation of norms in a biopolitical society brings into relief the condition of living in a normative universe, in which norms operate to confer and shape our bodily, ethical, social and political realities' (Mills 2011: 26–27). However, these norms are not simply imposed, but the norms of bodily life produce norms in relation to the surrounding social and material environment.

Lives that exceed or deviate from the vital or social norms elicit a response from biopolitical strategies of governance. In biopolitical societies, vital norms and social or political norms become interrelated in establishing legitimacy for action and intervention into lives. Much of the attention on Foucault's analyses of normalization has been on the social use of norms, however Canguilhem helps to articulate the interrelation between vital and social norms. For Canguilhem, a living being is normal by virtue of its capacity to internally produce new norms in relation to the environment. Thus disease is not abnormal but a new set of internally produced norms, albeit with a restricted capacity to produce new norms in relation to the surrounding social and material environment. The normativity of life – the capacity to produce new norms – suggests the possibility for transforming and resisting biopolitical strategies that attempt to capture or normalize life for social or political ends. However, both Mills and Mary Beth Mader warn against an overemphasis on vital norms (Mader 2007). The vital norms of bodies and the social norms of morality or politics are not equivalent. Vital norms are internal to the organism and can be observed while social norms are external and created (Pasquinelli 2015: 85). Social norms cannot simply be derived from or reduced to biological norms or biological norms from social. Social health is metaphorical in a way that physiological health is not (Mills 2013: 84). Foucault alludes to this irreducibility in *The Will to Knowledge* in saying that life always exceeds or escapes capture by techniques and strategies of governance. Canguilhem describes this excess as the capacity of error within life that gives rise to variance, difference and anomalous forms of life.

In reading Foucault through Canguilhem, Mills argues that normativity of life helps to re-think the relationship between politics and life. I suggest that this reading also demonstrates the way the 'urgent need' operates in the *dispositif* to incite and organize the tangle of power/knowledge/subjectivity in controlling life that exceeds the interrelation of vital and social norms. As Mills demonstrates, bodily life is not always reacting to politics, but life forces biopower to react. The excess or erring of life actively provokes biopower to create strategies to attempt to capture and standardize life, yet excess will constantly resist this capture. It is in this sense that the 'urgent need' provokes a reaction or response and activates the *dispositif* or enabling network.

### *Activating*

The provocation of sexuality or obesity for a response activates the *dispositif* and enables associated apparatuses. Without this activation the network would be dormant or unresponsive. For example, the *dispositif* of sexuality produces the subject of the homosexual or masturbating child, however it is the urgent need of fostering life and securing populations that mobilizes scientia sexualis in the pursuit of truth of sex and enables strategies of control and intervention (Foucault 1998).

In Foucault's analysis for *dispositif* of security in *Security, Territory, Population* (2007), he addresses the relationship between *dispositif* and apparatus in the context of population regulation. While the *dispositif* of security is the discursive and non-discursive network of power/knowledge that views the population as the source of wealth and strength of the state, the regulatory mechanisms of immigration procedures, for example, are the apparatus that exist within that *dispositif*. Thus the urgent need of securing the population activates the *dispositif* and enables apparatuses of control. It is in this sense that I argue lifestyle is a network that enables the governance of choices, activities and bodies. As a network comprised of lines of power, knowledge and subjectification, lifestyle responds to the 'urgent need' associated with bodies that exceed biological and social norms and activates a range of strategies to protect and secure the population.

An advantage of conceiving the *dispositif* as an enabling network opens the possibility for greater understanding of biopolitical strategies of governance. Liberalism and associated ideas of freedom and autonomy frame Singer and Callahan's arguments *for* intervention as well as Petr Skrabanek's and healthism arguments *against* such interventions. These approaches become entrenched in disputes over limits of government intervention and do not fully grasp the productive effects of power/knowledge and contingencies of subjects. This is particularly the case when trying to analyse amorphous and everyday ideas like 'lifestyle'. Constituted via multi-linear network of relations, practices and knowledges, lifestyle serves as a *dispositif* in which everyday lives, choices and bodies are made visible and governable. This is evident in response to the urgent need of securing the population from obesity. In the context of the obesity epidemic, the network of lifestyle enables health campaigns, experts, medical professionals, magazines, tape measures, smartphones and health policies as biopolitical strategies. Without the urgency of obesity or the entangled epidemiological and economic knowledges, these strategies would not be enabled or activated but remain dormant and unresponsive.

As has been outlined here, *dispositif* and apparatus are distinct concepts that need to be understood in relation to each other. To understand this relationship, I have argued that the *dispositif* can be conceived of as an enabling network that responds to an urgent need and in so doing activates apparatuses to govern and control that urgency. I contend that this understanding of *dispositif* and its relation to apparatus is useful in analysing biopolitical strategies that target and govern life.

## Biopolitics as art of government

Biopolitics is an increasingly popular concept that has been used to analyse the ways biological and social life is politically questioned in areas such as refugee policy, HIV prevention, border protection, ecological sustainability, genetic enhancement, and neoliberal economics. Thomas Lemke divides biopolitics along two main lines of thought (2011: 166). The first is via political philosophy, which has a focus on biopolitics as a 'mode of politics' that is connected with state racism of the late nineteenth and early twentieth centuries (Agamben 1998, Negri 2008, Esposito 2008). The second line of thought stems from the sociology of medicine and a focus on empirical questions regarding the 'substance of life' and the influence of biotechnology on the living body (Rose 2001, Petersen and Bunton 2002, Rabinow 2003, Cooper 2008). This line of interpretation emphasizes the 'seizure of life' through new developments in genetics that manipulate life at the molecular level. Although much of the biopolitical literature falls into one of these two groups, they are not incompatible interpretations. To analyse empirical interventions into life and the mode of politics driving such interventions requires both lines of thought. Critiques of the political control and governance of life requires the analysis of political processes, mechanisms and objectives, *and* attention to the material technologies, tools and discourses addressing the phenomena of life (Lemke 2011: 172).

Biopolitics and the *dispositifs* of security are particularly useful for addressing the enmeshed and entangled relations between the individual and the population. The problem of governing the individual and the population has a long history (Foucault 2000a, Mayes 2010). Do individuals comprise the whole? Or, do the interests of the whole outweigh those of the individual? In the liberal tradition, the population is commonly understood as the collection of individuals with rights based in universal laws and guaranteed by the state. The view of individuals separate from the population is presupposed by Mill's harm principle (1996: 78) and has been the object of much criticism (Berlin 2002: 201). The rise of the biological sciences from the eighteenth century challenges the idea that individuals can be separated from the whole. The idea that humans are not only part of the natural world, but biologically connected to each other destabilized the borders separating the individual-organism from the population-species. Foucault describes the significance of this transformation in the following way,

> Western man [*sic*] was gradually learning what it meant to be a living species in a living world, to have a body, conditions of existence, probabilities of life, an individual and collective welfare, forces that could be modified, and a space in which they could be distributed in an optimal manner.
>
> (Foucault 1998: 142)

No longer conceived as a collection of distinct individuals with interests and rights, the population comes to be understood as 'a set of natural phenomena with its own immanent laws of transformation and movement' (Blencowe 2012: 63).

The spaces of distribution, immanent laws and probabilities become both the object and result of epidemiological, statistical and economic knowledge (Foucault 2007: 350ff). These new fields of knowledge enable the regulation and government of life through town planning, insurance schemes and vaccination programmes (Garland 2014). 'For the first time', writes Foucault, 'biological existence was reflected in political existence' (1998: 142). The introduction of life into politics has led to the emergence of biopolitics.

The objective of biopolitics is to 'fabricate, organize, and plan' the milieu in order to regulate and govern the life of the population (Foucault 2007: 21).[4] Biopolitics seeks to both derive knowledge from and intervene in the environmental phenomena that affect the population. The concept of obesogenic environment is an example of the production of knowledge to enable biopolitical intervention into the lives of individuals for the security of the population. The obesogenic environment is defined in public health as 'the sum of influences that the surroundings, opportunities, or conditions of life have on promoting obesity in individuals or populations' (Swinburn, Egger, and Raza 1999: 564). Epidemiological research is used to characterize activities such as playing computer games, microwave dinners, sugary drinks, feeding infants formula, alcohol consumption, or mode of transportation as indicators not only of individual health or disease risk, but threats to population security (Holland et al. 2011, Lupton 2004, Gard and Wright 2005). These are the activities and choices that Callahan wants governments to intervene in and alter. As such, these choices and behaviours are no longer benign and inconspicuous but are politicized as targets for a specific form of intervention that organize the natural and created environment (Bell, McNaughton, and Salmon 2009). Individuals engaging in these practices are called to take responsibility for themselves in order to mitigate the possibility of catastrophe at the level of population. From this perspective, the bioethics of Singer and Callahan can be understood as a biopolitics.

### *Governing with norms*

The establishment of norms through health statistics and population data enables the governance of individuals and regulation of the population. The threat associated with health-related activities of individuals is communicated and understood through the scalar logic of the norm. Mary Beth Mader writes 'Law breaks a citizenry into obedient and disobedient subjects. But the norm has a refined, multiplied classification scheme that operates on a scalar model' (2007: 9). While the law operates with a binary logic of obedient/disobedient or legitimate/illegitimate, the norm operates with a scalar logic. The individual is measured in relation to others and the population along a continuous scale. The scalar logic of the norm implies new and different forms of governance and control. Unlike the law, this enables what Foucault calls '*continuous regulatory and corrective mechanisms*' (1998: 144). Systems of surveillance are introduced and produce 'an incessant visibility, a permanent classification of individuals, the creation of a hierarchy of qualifying, establishing limits, providing diagnostics'

(Foucault 1996: 197). Through the norm the individual is weighed, measured, evaluated and judged.

The process of governing individuals and regulating populations is referred to as normalization. In an interview Foucault remarks that 'Medical power is at the heart of a society of normalization' (1996: 1997). This medical power or biopower is not exclusively concerned with disciplining the individual-body or regulating the population-species. Rather the circulation of the norm in regulatory and disciplinary technologies normalizes society by enabling the biopolitical seizure of the individual-body and population-species, as well as the spaces between (Foucault 2004: 253). According to Cressida Heyes, normalization operates through 'macro-territories' such as health policy and socioeconomics that regulate the population as well as 'micro-territories' of 'the body's surface and interior' that discipline the individual (2007: 17). These two levels of normalization are significant in the discussion of lifestyle shaping the individual and securing the population through norms of health. The circulation of the norm occurs through individuals 'voluntarily' embracing norms, and populations recognizing and affirming them. As is elaborated in the following chapters, norms are not applied against the will of an individual but chosen, as a flock chooses to follow the path created by the shepherd to fertile pastures (Mayes 2010). Norms are applied to and through the discipline of individuals and regulation of populations. Thus it is important that a conception of biopolitics accounts for the function of both.

Foucault argues that the 'norm is something that can be applied to both a body one wishes to discipline and a population one wishes to regularize' (2004: 253). Yet without reference to the population, norms could not operate on the individual, as there would be no point of reference to which the individual had to conform. Discussing the relation between the norm and the individual in *Discipline and Punish*, François Ewald notes that 'the norm participates in this logic of individualization while also serving as the force that joins together the individuals created by discipline and allows them to communicate with one another' (1990: 141). The individual is produced through norms that measure the individual in relation to the common standard of the population. The norm ties and entangles individuals together and trains the individual-body to conform to the population-species. When the individual conforms to the norm, they are embraced and absorbed into the population. However, when the individual's body, attitude, or practice deviates from the norm they are made visible and exposed to the population. The norm makes the deviant or abnormal individual conspicuous or stigmatized and attracts the attention of the 'pastor' or mechanism of security. The conspicuousness of the individual-body in relation to the population-species leads to increasingly direct forms of governance.

### Care and violence

In his influential study of stigma, Erving Goffman writes that 'By definition, of course, we believe the person with a stigma is not quite human' (1983: 2). Goffman is not interested in the stigmatising attribute per se, but the social relations, or what

Foucault would call the *dispositif*, that transforms a behaviour or characteristic into a stigma that disqualifies and discredits the bearer from full participation in the community. The disqualifying and discrediting effect of stigma can expose the bearer to direct violence or at least place them in a context where they are no longer part of the community that is cared for and secured from violence. The dynamics of care and violence can be seen in Foucault's analyses of pastoral power – a precursor to biopower and governmentality.

In a 1982 seminar Foucault described governmentality as the combination of technologies of domination and technologies of the self (Foucault 2000b), that is, a rationality of government that dovetails discipline and control with ethics and freedom. However, Foucault's *Security, Territory, Population* lectures provide a more historical and detailed analysis. Guiding Foucault's inquiry into the development of biopolitics and history of governmentality is the question: under what conditions did the idea of governing a people, as distinct from a territory, emerge? Foucault suggests that it was the Judeo-Christian pastor, not the Greek or Roman statesman, who first sought to govern a people rather than a territory or state.[5] According to Foucault, the pastor serves as the 'embryonic point of the governmentality whose entry into politics…marks the threshold of the modern state' (2007: 165). The power of the pastor is directed towards the flock or people for their benefit – 'pastoral power is a power of care' for the salvation of the flock (2007: 126). There are four features of the pastor: he governs a people rather than a territory; he governs for the benefit of the people rather than something external to it like a state; he is equally concerned for the one and the many; and he needs to know the consciences of the flock. This fourth feature is particularly significant for Foucault as it introduces the technique of confession (Taylor 2008, Mayes 2009). Confession seeks to uncover knowledge of the conscience, the inside of people's minds, thereby allowing the governance of will and desire (Foucault 1983: 214). Foucault suggests that these features of pastoral power were introduced to Western practices of governance through the Christian Church, which has influenced both institutional and non-institutional practices of government. With Christianity begins 'the real history of the pastorate as the source of a specific type of power over men, as a model and matrix of procedures for the government of men' (Foucault 2007: 147–148).

The intermingling of violence and care forms a key feature of pastoral power and influences the conception of biopolitics as an art of government. Foucault outlines the caring and protecting features of the shepherd/pastor in the Exodus narrative. God is the shepherd that leads his people out of slavery and into the Promised Land (after a few detours in the desert). However, in his recounting of the narrative, Foucault neglects the violence of the plagues and the drowning of Pharaoh's army that are attributed to the hand of God. The violent events of plagues and mass drowning made the exodus possible. Both the demonstration of violence and the ability to protect confirmed God's role as the pastor who would lead the people to fertile pastures (Exodus 14:11–12).

The care of the pastor can manifest as violence or exclusion of those that threaten or destabilize the flock.[6] This feature of care and violence can be seen in other biblical shepherd-pastors from David through to Paul (Mayes 2010). This genealogy of pastoral care and violence provides the backstory for modern biopolitics as a 'power to *foster* life or *disallow* it to the point of death' (Foucault 1998: 138). It also speaks to the debates over the foundation of biopolitics. Mika Ojakangas argues that the love and care of the pastor forms the hidden foundation of biopolitics. Arguing against Giorgio Agamben, Ojakangas suggests that it is not bare life's exposure 'to an unconditional threat of death, but that the *care of "all living"*' that provides 'the foundation of bio-power' (2005: 6). Following Foucault, Ojakangas traces modern forms of government back through the Christian tradition of the West suggesting that 'the origin of bio-political rationality can be found in the Judeo-Christian tradition of pastoral power' (2005: 19). While Agamben is criticized for turning biopolitics into a 'thanatopolitics' (a politics of death), Ojakangas takes biopolitics in the opposite direction towards an 'agape-politics' (a politics of love). Ojakangas argues that it 'is precisely care, the Christian power of love (*agape*), as the opposite of all violence that is at issue in biopower' (2005: 20). For Ojakangas, the care found in pastoral power provides the foundation for biopolitics, rather than the violence directed toward bare life as it is for Agamben. However, I suggest that in pastoral power, and especially in the logic of sacrifice that it entails, *both* care and violence operate in and inform biopolitics.

The nexus of violence and care contained within the Christian pastorate extends into the 'model and matrix of procedures for the government of men' (Foucault 2007: 147–148). The violence of exclusion as the expression of pastoral care for the flock establishes and maintains the order of the community. This violence serves to maintain an order and a stable state of affairs that protects and secures the population, while excluding those elements that threaten this security. The dynamics of violence and care operate through the biopolitical *dispositif* of lifestyle to structure social relations and identify subjects to be cared for and those that can be excluded from care.[7]

The care for the life of the population may well require the exclusion of those that threaten its health, safety, and security. This can be seen in past and present practices of quarantine (Bashford 2006, Elbe 2010). However, I contend that lifestyle also serves as a mechanism of care and exclusion, but in a subtler fashion. Lifestyle works as a network that enables the isolation of the bodies and choices of individuals and governs them in relation to the population. If the individual is able to adopt and develop a 'healthy lifestyle' they remain inconspicuously nestled within the secured population. However, when the bodies and choices of individuals deviate from norms associated with the health of the population, then the relation between care and violence becomes ambiguous. As will be demonstrated, there are attempts to entice the individual towards a 'healthy lifestyle' but repeated failure or assumed inability to make such a transition exposes the individual to an environment with minimal welfare guarantees.

The focus on purported healthy choices highlights these dynamics. Both Singer and Callahan argued that individuals who *appear* to deviate from bodily norms can be justifiably coerced and should have to bear the economic burdens associated with their appearance. The population is harmed by the individual and the need to care for the whole justifies the exclusion and stigmatization of the few that threaten it. According to this account, the individual is responsible for their own welfare and cannot expect social welfare provision to protect them from their own choices or the inability to choose due social and structural factors. Furthermore, the *appearance* as a threat can also provoke a response from disciplinary mechanisms.

Unlike Skrabenak and the healthism critique that argues against lifestyle health programmes because they allow governments to intervene in the lives of individuals, the biopolitical critique of lifestyle outlined in this book argues that lifestyle is problematic because it recodes freedom and social relations in a way that isolates individuals and makes them responsible for outcomes that are not necessarily within their control. The healthism critique, like the lifestyle network, tends to ignore or minimize the broader structural issues that disproportionality effect historically marginalized individuals and groups within populations. In this sense, the lifestyle interventions are not bioethical, but biopolitical. And as such, there needs to be a biopolitical critique. The aim of this book is to provide such a critique.

## Notes

1　Some readers may object that *Project Syndicate* has a popular-level audience requiring short opinion pieces and it is therefore inappropriate to critique Singer's article on scholarly grounds. I reject such objections. The push for academics to engage wider audiences is laudable, however by circumventing processes such as peer-review it becomes even more important for authors and scholarly communities to ensure ethical and research standards are maintained.

2　A similar debate occurs over the translation of Guattari's use of *dispositif* and its relation to *agencement*. See (Guattari 2000: 76, n17).

3　For a useful analysis of the differences between Agamben's and Foucault's genealogies of *dispositif* see (Pasquinelli 2015).

4　A further element that cannot be addressed here is the environment in which living beings exist. Georges Canguilhem, Foucault's dissertation supervisor and colleague, notes the dynamic and active relationship between the living being and environment, arguing that 'the environment of the living being is also the work of the living being who chooses to shield himself from or submit himself to certain influences' (Canguilhem 2007: 179). The living being is in a dynamic bond with both the 'natural' and the 'created' environment.

5　Foucault argues that for the Greek ruler the 'object or target of government is the city-state in its substantial reality, its unity, and its possible survival or disappearance' (Foucault 2007: 123). Thus Foucault concludes that the idea of governing people was not a Greek idea (or Roman); rather, this idea has its roots in the pre-Christian East, first with the Egyptians and Assyrians but primarily with the Hebrews, and then with the early Christians.

6　In Hebrew texts, this theme is evident in the Exodus account referred to above, but also in the narrative of David, the first shepherd of men, in which the shepherd's ability to kill predators becomes a significant feature. It is David's experience of caring for the flock that required him to kill for the flock – lions and wolves in his literal role as a shepherd – but this experience is then mirrored by his care for Israel

that requires him to kill Goliath for the protection and security of Israel. See 1 Samuel 17: 34–54 and 2 Samuel 5: 2.
7  For a further discussion of the role of violence, sacrifice and the community see (Girard 2005).

# References

Agamben, Giorgio. 1998. *Homo Sacer: Sovereign Power and Bare Life*. Translated by Daniel Heller-Roazen. Stanford, CA: Stanford University Press.

Agamben, Giorgio. 2009. *What is an Apparatus? And Other Essays*. Translated by David Kishik and Stefan Pedatella. Edited by Werner Hamacher. Stanford, CA: Stanford University Press.

Anonymous. 2013. "North America weight loss/obesity management market worth $139.5 billion by 2017" *Wall Street Journal* [cited July 13 2013]. Available from http://online. wsj.com/article/PR-CO-20130708-903278.html.

Bashford, Alison. 2006. "Global biopolitics and the history of world health." *History of the Human Sciences* 19 (1):67–88. doi: 10.1177/0952695106062148.

Baum, Fran. 2011. "From Norm to Eric: avoiding lifestyle drift in Australian health policy." *Australian and New Zealand Journal of Public Health* 35 (5):404–406. doi: 10.1111/j.1753-6405.2011.00756.x.

Baum, Frances Elaine, and David M. Sanders. 2011. "Ottawa 25 years on: a more radical agenda for health equity is still required." *Health Promotion International* 26 (suppl 2):ii253–ii257. doi: 10.1093/heapro/dar078.

Bell, Kirsten, Darlene McNaughton, and Amy Salmon. 2009. "Medicine, morality and mothering: public health discourses on foetal alcohol exposure, smoking around children and childhood overnutrition." *Critical Public Health* 19 (2):155–170. doi: 10.1080/09581590802385664.

Berlin, Isaiah. 2002. "Two Concepts of Liberty." In *Liberty*, edited by Henry Hardy. New York: Oxford University Press.

Bishop, Jeffrey P., and Fabrice Jotterand. 2006. "Bioethics as biopolitics." *Journal of Medicine and Philosophy* 31 (3):205–212.

Blencowe, Claire. 2012. *Biopolitical Experience: Foucault, Power and Positive Critique*, Basingstoke: Palgrave Macmillan.

Brownell, Kelly D., Rogan Kersh, David S. Ludwig, Robert C. Post, Rebecca M. Puhl, Marlene B. Schwartz, and Walter C. Willett. 2010. "Personal responsibility and obesity: a constructive approach to a controversial issue." *Health Affairs* 29 (3):379–387. doi: 10.1377/hlthaff.2009.0739.

Bussolini, Jeffrey. 2010. "What is a Dispositive?" *Foucault Studies* 10:85–107.

Callahan, Daniel. 2013. "Obesity: chasing an elusive epidemic." *Hastings Center Report* 43 (1):34–40.

Cameron, David. 2010. "Return to responsibility." *The Guardian*, 28 February.

Canguilhem, Georges. 2007. *The Normal and the Pathological* Translated by Carolyn R. Fawcett. New York: Zone Books.

Cheng, Jennifer K. 2012. "Confronting the social determinants of health—obesity, neglect, and inequity." *New England Journal of Medicine* 367 (21):1976–1977.

Cooper, Melinda. 2008. *Life as Surplus: Biotechnology and Capitalism in the Neoliberal Era*. Seattle, WA: University of Washington Press.

Dawson, Angus. 2010. "The future of bioethics: three dogmas and a cup of hemlock." *Bioethics* 24 (5):218–225.

Deleuze, Gilles. 1992. "What is a *dispositif?*" In *Michel Foucault, Philosopher*, edited by Timothy J. Armstrong. New York: Routledge.

Dreyfus, Hubert, and Paul Rabinow. 1983. *Michel Foucault: Beyond Structuralism and Hermeneutics*. 2nd edn. Chicago, IL: University of Chicago Press.

Elbe, Stefan. 2010. *Security and Global Health*. Cambridge: Polity.

Elster, Jon. 1982. "The case for methodological individualism." *Theory and Society* 11 (4):453–482. doi: 10.2307/657101.

Esposito, Roberto. 2008. *Bios: Biopolitics and Philosophy*. Translated by Timothy Campbell. Edited by Cary Wolfe. Vol. 4 of *Posthumanities*. Minneapolis, MN: University of Minnesota Press.

Ewald, François. 1990. "Norms, discipline, and the law." *Representations* 30:138–161.

Flegal, Katherine M., Brian K. Kit, Heather Orpana, and Barry I. Graubard. 2013. "Association of all-cause mortality with overweight and obesity using standard body mass index categories: A systematic review and meta-analysis." *JAMA* 309 (1):71–82. doi: 10.1001/jama.2012.113905.

Foucault, Michel. 1980. "The confession of the flesh." In *Power/Knowledge: Selected Interviews and Other Writings*, edited by Colin Gordon. New York: Pantheon Books.

Foucault, Michel. 1983. "Afterword: the subject and power." In *Michel Foucault: Beyond Structuralism and Hermeneutics*, edited by Hubert Dreyfus and Paul Rabinow. Chicago, IL: University of Chicago Press.

Foucault, Michel. 1996. "The social extension of the norm." In *Foucault Live: Collected Interviews, 1961–1984*, edited by Sylvère Lotringer. New York: Semiotext(e).

Foucault, Michel. 1998. *The Will to Knowledge: The History of Sexuality Volume 1*. Translated by Robert Hurley. Harmondsworth: Penguin Books.

Foucault, Michel. 2000a. "'Omnes et singulatim': toward a critique of political reason." In *Power* edited by James D Faubion. Harmondsworth: Penguin.

Foucault, Michel. 2000b. "Technologies of the self." In *Ethics: Subjectivity and Truth*, edited by Paul Rabinow. Harmondsworth: Penguin.

Foucault, Michel. 2003. *Abnormal: Lectures at the Collège de France 1974–1975*. Translated by Graham Burchell. Edited by Arnold I. Davidson. New York: Picador.

Foucault, Michel. 2004. *Society Must Be Defended: Lectures at the Collège de France 1975–76* Translated by David Macey. Edited by Arnold I. Davidson. Harmondsworth: Penguin.

Foucault, Michel. 2005. *The Hermeneutics of the Subject: Lectures at the Collège de France 1981–1982*. Translated by Graham Burchell. Edited by Arnold I. Davidson. New York: Picador.

Foucault, Michel. 2006. *Psychiatric Power: Lectures at the Collège de France, 1973–74*. New York: Palgrave Macmillan.

Foucault, Michel. 2007. *Security, Territory, Population: Lectures at the Collège de France 1977–78*. Translated by Graham Burchell. Edited by Arnold I. Davidson. New York: Palgrave Macmillan.

Galer-Unti, Regina A. 2013. "Social epidemiology." In *Oxford Bibliographies Online: Public Health*, edited by David V. McQueen. Oxford University Press.

Gard, Michael, and Jan Wright. 2005. *The Obesity Epidemic: Science, Morality and Ideology*. New York: Routledge.

Garland, David. 2014. "The welfare state: a fundamental dimension of modern government." *European Journal of Sociology* 55 (3):327–364.

Girard, Rene. 2005. *Violence and the Sacred*. Translated by Patrick Gregory. New York: Continuum.

Goffman, Erving. 1983. *Stigma: Notes on the Management of Spoiled Identity*. New York: Touchstone.

Goldberg, Daniel S. 2012. "Social justice, health inequalities and methodological individualism in US health promotion." *Public Health Ethics* 5 (2):104–115.

Goldberg, Daniel S., and Rebecca M. Puhl. 2013. "Obesity stigma: a failed and ethically dubious strategy." *Hastings Center Report* 43 (3):5–6.

Guattari, Felix. 2000. *The Three Ecologies*. Translated by Ian Pindar and Paul Sutton. London: Athlone Press.

Guttman, Nurit, and Charles T. Salmon. 2004. "Guilt, fear, stigma and knowledge gaps: ethical issues in public health communication interventions." *Bioethics* 18 (6):531–552.

Heyes, Cressida J. 2007. *Self Transformations: Foucault, Ethics, and Normalized Bodies*. New York: Oxford University Press.

Holland, Kate E., R. Warwick Blood, Samantha I. Thomas, Sophie Lewis, Paul A. Komesaroff, and David J. Castle. 2011. "'Our girth is plain to see': An analysis of newspaper coverage of Australia's future 'fat bomb'." *Health, Risk & Society* 13 (1):31–46. doi: 10.1080/13698575.2010.540648.

Kassirer, Jerome P., and Marcia Angell. 1998. "Losing weight – an ill-fated New Year's resolution." *New England Journal of Medicine* 338 (1):52.

Lemke, Thomas. 2011. "Beyond Foucault: From biopolitics to the government of life." In *Governmentality: Current Issues and Future Challenges*, edited by Ulrich Bröckling, Susanne Krasmann and Thomas Lemke. New York: Routledge.

Lupton, Deborah. 2004. "'A grim health future': food risks in Sydney press." *Health, Risk and Society* 6 (2):187–200.

Mader, Mary Beth. 2007. "Foucault and social measure." *Journal of French Philosophy* 17 (1). DOI: http://dx.doi.org/10.5195/jffp.2007.203

Marmot, Michael. 2010. *Fair Society, Healthy Lives*, The Marmot Review. Available from http://www.instituteofhealthequity.org/Content/FileManager/pdf/fairsocietyhealthylives.pdf.

Mayes, Christopher. 2009. "Pastoral power and the confessing subject in patient-centred communication." *Journal of Bioethical Inquiry* 6 (4):483–493.

Mayes, Christopher. 2010. "The violence of care: an analysis of Foucault's pastor." *Journal of Cultural and Religious Theory* 11 (1):111–126.

Mill, John Stuart. 1996. *Utilitarianism, On Liberty, Considerations on Representative Government*. London: Everyman.

Mills, Catherine. 2010. "Continental philosophy and bioethics." *Journal of Bioethical Inquiry* 7 (2):145–148.

Mills, Catherine. 2011. *Futures of Reproduction: Bioethics and Biopolitics*. New York: Springer.

Mills, Catherine. 2013. "Biopolitical life." In *Foucault, Biopolitics and Governmentality*, edited by Jakob Nilsson and Sven-Olov Wallenstein. Stockholm: Södertörn Philosophical Studies.

Negri, Antonio. 2008. *Empire and Beyond*. Translated by Ed Emery. Cambridge: Polity Press.

O'Hara, Lily, and Jane Gregg. 2006. "The war on obesity: a social determinant of health." *Health Promotion Journal of Australia* 17 (3): 260–263.

Ojakangas, Mika. 2005. "Impossible dialogue on bio-power: Agamben and Foucault." *Foucault Studies* 2:5–28.

Pasquinelli, Matteo. 2015. "What an apparatus is not: on the archeology of the norm in Foucault, Canguilhem and Goldstein." *Parrhesia* 22:79–89.

Petersen, Alan, and Robin Bunton. 2002. *The New Genetics and the Public's Health*. London: Routledge.

Public Health Association of Australia. 2010. *Promoting Healthy Weight Policy*. edited by Andrea Begley and Christina Pollard, Sydney: Public Health Association of Australia.

Rabinow, Paul. 2003. *Anthropos Today: Reflections on Modern Equipment*. Princeton, NJ: Princeton University Press.

Rose, Nikolas. 2001. "The politics of life itself." *Theory, Culture & Society* 18 (6):1–30.

Rudd Center. 2013. "Yale Rudd Center For Food Policy and Obesity". [cited September 23 2013]. Available from http://www.yaleruddcenter.org/.

Singer, Peter. 2012. "Weigh more, pay more". *Project Syndicate: A World of Ideas*, http://www.project-syndicate.org/commentary/weigh-more-pay-more.

Swinburn, Boyd, Garry Egger, and Fezeela Raza. 1999. "Dissecting obesogenic environments: the development and application of a framework for identifying and prioritizing environmental interventions for obesity." *Preventive Medicine* 29 (6):563–570. doi: 10.1006/pmed.1999.0585.

Taylor, Chloë. 2008. *The Culture of Confession from Augustine to Foucault: A Genealogy of the 'Confessing Animal'*, London: Routledge.

Venkatapuram, Sridhar, Ruth Bell, and Michael Marmot. 2010. "The right to sutures: social epidemiology, human rights, and social justice." *Health and Human Rights: An International Journal* 12 (2): 3–16.

Veyne, Paul. 2010. *Foucault: His Thought, His Character*. Cambridge: John Wiley & Sons.

# 2   Lifestyle as politics
## Choice and responsibility

> Freedom is a tenable objective only for responsible individuals...We cannot categorically reject paternalism for those whom we consider as not responsible.
> Milton and Rose Friedman (1980: 32–33)

Choice and responsibility are persistent themes in the rhetoric describing the global rise in obesity. Narratives of excess and a decline of traditional virtues of self-control are used to lament the erosion of responsible choice and subsequent increase in obesity. Although the individual is the main actor, these narratives are not primarily concerned with individual wellbeing but the societal consequences of individuals' choices. The aggregation of irresponsible choices is used to establish links between obesity and national or global security in biopolitical societies (Elbe 2010: 132ff). Perhaps the most incendiary example of this rhetoric comes from the former US Surgeon General Richard Carmona, who compared obesity in the United States to domestic terrorism, warning that unless there is action 'the magnitude of the dilemma will dwarf 9-11 or any other terrorist attempt' (Carmona 2003). Comparing obesity with terrorism is a common motif in anti-obesity rhetoric, especially since 9/11 (O'Hara and Gregg 2006, Gard 2007, Biltekoff 2007). However, economic disaster resulting from increasing health expenditure is a primary concern used to justify new health policies.

The emphasis on economic costs extends obesity from an individual moral or medical problem to a threat to national security. The former Australian Minister for Health and Ageing, Nicola Roxon, evidences this shift in stating, 'the cost of obesity is not only a personal one but also a huge drain on the nation's economy' (2011). Obesity is estimated to have an annual cost of $58 billion in Australia, $147 billion in the US, and between $30 billion and $125 billion in the UK.[1] Although questions can be raised over the accuracy of these calculations, my interest is in the use of such figures in the lifestyle network. Using economics to link obese individuals with population security creates an avenue that justifies the modification of services and the deployment of interventions to change individual choice.

During his 2015 campaign for re-election, British Prime Minister David Cameron announced that his government might cut sick benefits for people who are obese and do not lose weight (Boseley 2015). The rationale behind this strategy is that obese people can lose weight simply by making responsible choices. According to Cameron, people are not making these choices because they have become dependent on welfare benefits, and it is in their own interest, and the nation's, to become independent. This rationale is widespread among governments in the US, the UK and Australia, where it is believed that 'ultimately individuals must take responsibility for their own health, including their weight' (Australia Parliament House of Representatives Standing Committee on Health and Ageing 2009: 119).[2] The strategies undertaken in these countries are not identical. However, the idea that the individual is ultimately responsible for their health and its impact on society reflects the logics of neoliberalism that informed the practices of governance in Australia, the US, the UK and elsewhere since the 1970s.

Neoliberal rationality of governance provides the political context to understand the role of individual choice in the lifestyle network. The influence of neoliberal ideas on economics and politics contributes to a network that enables the governance of individual choice and the subjectification of individuals as responsible for the security of the population. Recalling Deleuze's summary of a *dispositif* as 'lines of visibility and enunciation, lines of force [and] lines of subjectification' (1992: 162), I argue that neoliberal economics and public health rhetoric in Australia, the US, and the UK produce 'lines of visibility and enunciation', making it possible to 'see' and 'speak' of the choices of the individual as issues of security. Drawing further on Foucault's lectures on biopolitics and neoliberal governmentality, this chapter traces the way the lifestyle network problematizes individual choice as a security issue requiring biopolitical governance. I identify three tensions in the biopolitical governance of obesity: present and future, freedom and security, individual and population. I then demonstrate how these operate to care for 'healthy subjects' and exclude 'irresponsible subjects'. I conclude the chapter with an analysis of techniques of governance employed in the Australian government's *Measure Up* social marketing campaign. I use this campaign as an illustration of neoliberal health policies that seek to govern individual choice and bodies toward norms of health as a means of securing the population.

## Neoliberal Governmentality: the individual, state-phobia and responsibility

Neoliberalism is ill-defined, and often used in a pejorative sense to designate an approach to governance that values 'the superiority of individualized, market-based competition over other modes of organization' (Mudge 2008: 706–707). There is no consensus definition of neoliberalism, however common themes include privatization of social services (Alejandro Leal 2007, Rowe and Frewer 2005), individual choice and responsibility (Binkley 2009, Ilcan 2009, Dilts 2011), deregulation of private enterprise (Harvey 2009, Lazzarato 2009) and the use of market mechanisms to govern society (Mudge 2008, Mirowski 2009).

A common complaint about neoliberal approaches to government is the tendency for market logics to extend into areas previously considered outside of the purview of the market (Zoller 2009, Terris 1999). Addressing the expansion of neoliberalism in his *The Birth of Biopolitics 1978–79* course, Foucault notes 'we are seeing the economic policies of all the developed countries, but also their social policies, as well as their cultural and educational policies, being orientated in these terms' (2008: 232). Whether for education, healthcare, environmental conservation, transport or energy infrastructure, proponents of neoliberal policies claim that the free market provides the most efficient mode of organization, while also guaranteeing the freedom of individuals (Biancardi 2003: 187–188).

In mapping the genealogy of governmentality from the early Christians to neoliberal economic theories in America and West Germany, Foucault argues that the law recedes to the background while the norm becomes the prominent mode of governance. Foucault contends that conducting the conduct of individuals and populations 'is not a matter of imposing a law', but of employing tactics to arrange 'things so that this or that end may be achieved through a certain number of means' (Foucault 2007: 99). Under neoliberal governance, self-regulation, voluntary codes of conduct and incentives are used in preference to direct interventions (McNay 2009: 63). This is not to imply that neoliberalism makes the law and State redundant. As is clear from observing any 'actually existing' neoliberal government, the disciplinary and coercive force of the police and law is ready to intervene if needed (Cahill 2014: 26, Whyte 2014: 218ff). The law is not superfluous, but it is no longer the primary mode of government: norms and incentives operate in the shadow of the law. Norms and incentives become the tools for regulating and governing the individual and the population as they operate continually, yet at a distance, to entice individuals towards the objectives of security (Rose, O'Malley, and Valverde 2006).

In *The Birth of Biopolitics*, Foucault had intended to continue the investigation of biopolitics that he began in the *Security, Territory, Population* lectures. However, he remarks that to understand biopolitics he first needs to know the governmental regime of liberalism (Foucault 2008: 22). To the confusion and frustration of some readers, *The Birth of Biopolitics* lectures do not address biopolitics as such, but are devoted to a detailed analysis of the interaction between Keynesian liberalism, Ordoliberalism and the emergence of Neoliberalism.[3] A key effect of neoliberal rationality is the transformation of the relationships between the individual, the State and the economy. Central to this change is the idea that freedoms are best guaranteed and protected by the free market. Rather than the State governing the functioning of the economy or the provision of social services, the free market is considered an efficient and emancipatory tool of governance.

Margaret Thatcher is a key figure in the political history of neoliberalism. Although she is not responsible for the political and economic theories, she did popularize and implement policies that altered the relationship between the State, the economy and the individual. In an interview during her first term as Prime Minister, Thatcher expressed irritation at the way post-war politics privileged collective society over what she referred to as 'personal society'. When asked how she intended to achieve the shift from the collective to the individual, Thatcher replied, 'Economics are the method; the object is to change the heart and soul'

(Butt 1981). Neoliberal governmentality uses norms and economics to shape the 'heart and soul' from a distance, and in so doing transforms the focus of Western politics from the collective to the individual.

Although political leaders like Thatcher and Ronald Reagan have come to symbolize neoliberalism, Friedrich von Hayek is arguably its chief intellectual force. Hayek has had a profound, if misunderstood, influence on the political and economic theory of neoliberalism, particularly the emphasis on the market as protector of individual freedoms (Hayek 2001: 104). In 1947, Hayek, along with Ludwig von Mises and Milton Friedman, formed the Mont Pelerin Society to promote the ideas of freedom of the individual and human dignity. The Mont Pelerin Society believed that these ideas were threatened by a 'decline of belief in private property and the competitive market; for without the diffused power and initiative associated with these institutions it is difficult to imagine a society in which freedom may be effectively preserved' (1947). It argued that the free market and private property are the essential institutions that ensure a free society. Although Hayek and Friedman had different perspectives on the role of the state (Cahill 2014: 37ff), a general consensus from the Mont Pelerin Society was that the primary roles of the State are to enforce laws that protect private property, and to enable the market to operate freely, unhindered by fraud or outside influence. Aside from a limited role for the State, the two principal ideas of Hayek's neoliberalism are that centralized planning poses a threat to freedom, and that the market, as a self-organizing system in which individuals freely choose to participate, is best suited to guarantee freedom (Hayek 2001, Devine 2007: 41). The logic of neoliberalism maintains that the free market maximizes wealth and security for individuals and society, while the State, trade unions or other forms of collective planning threaten freedom.

### Free to be responsible

Freedom is perhaps the most distinctive feature of the neoliberal framework. Yet as Isaiah Berlin wrote, freedom 'is a term whose meaning is so porous that there is little interpretation that it seems able to resist' (2002: 168). According to Berlin, 'conceptions of freedom directly derive from views of what constitutes a self' (2002: 181). To properly understand neoliberal governmentality it is important to be clear about the concept of freedom and the subject that is presupposed. Neoliberal freedom is not equivalent to the *laissez-faire* liberal sense of freedom that presupposes a fixed subject that must be let alone. According to Mirowski, neoliberalism presupposes the self as rational and motivated by 'ineffable self-interest, striving to improve their lot in life by engaging in market exchange' (2009: 437). However, what is considered rational and self-interested is not fixed or stable, but open to modification and in continual flux (Gordon 1991: 43). Thus the neoliberal subject is not simply 'free to choose' as Milton Friedman and Rose Friedman suggest (1980), but is surrounded by biopolitical mechanisms that normalize, control and style the choosing subject toward healthy, rational and responsible choice. I elaborate on this below, however for the moment it is sufficient to note that freedom in neoliberal thought is 'recoded and heavily edited' (Mirowski 2009: 437).

Freedom is used in neoliberal governmentality to establish new relationships between individuals, the market and government. People are governed *through* freedom, not in conflict with freedom. Nikolas Rose writes that freedom has 'come to provide the grounds upon which government must enact its practices for the conduct of conduct' (1999: 11). The emphasis on freedom as the guarantor of the practices of government – the 'conduct of conduct' – mobilizes freedom as 'a cardinal feature of neoliberal theory' (Harvey 2009: 7). However, conceiving freedom as the 'grounds upon which government must enact its practices' entails an erosion of collective protection. The free participation of individuals in a competitive market does not necessarily provide security against one's own 'irresponsible' choice. Neither does it necessarily provide safeguards for those with limited or no choice: neoliberal governmentality drives the logic of the market 'into the fine grain of everyday life' so that public services, working conditions, and social security can be 're-coded and re-organised as choices' (Pusey 2010: 139–140). Basic needs such as health, education or shelter are not immune from this process of reframing collective public policy as individual choice and responsibility.

It is worth noting that a neoliberal state has not simply displaced the post-war welfare state, where the latter represents the 'good old days' of communal concern, while the former is a bleak dystopia of competition. Such narratives are commonplace, yet they ignore the disciplinary features of post-war state institutions that gave 'greater security' but also marginalized and excluded (Foucault 2000). They also ignore the continued provision of social services by neoliberal governments. Though they have made sizable funding cuts and eroded public institutions (Pollock 2005), David Garland notes, 'there is no state in the industrialized world that lacks a developed welfare state apparatus or which does not devote a significant fraction of its budget to social expenditures' (2014: 332). 'Welfare' is still part of neoliberal governmentality, but like freedom it is recoded and edited. Importantly, this has involved the narrowing of *who the welfare is for* and the means by which it is determined. Neoliberal governmentality and the emphasis on individual freedom and responsible choice intensify and narrow the community that is cared for and secured. The lifestyle network of neoliberal governmentality produces a self-selecting community comprised of responsible subjects. Those who are unable to participate in the community are self-excluded by their irresponsible choices. Under this logic, to expect the State to intervene and redress the irresponsible choices of individuals not only dehumanizes them, but threatens the whole.[4]

### Fear of the state

In his inaugural Presidential address, Ronald Reagan quipped, 'in this present crisis, government is not the solution to our problem, Government is the problem' (Howell and Ingham 2001: 329). Reagan's rhetoric has continued to influence a deep suspicion that government programmes and collective planning external to the market are incapable of directing or protecting either individual or collective interests. Echoes of Reagan's suspicion of government can still be heard in election campaigns and major policy announcements throughout Western liberal democracies. In the US, for instance, concerns over the federal government's role in taxation or unemployment, as well as the debates over the *Patient Protection and Affordable Care Act* (better known as Obamacare) exemplify the fear of 'Big Government' interference that will take away individual freedom. This is perhaps most colourfully expressed in H. Von Bulow's self-published *How Fascism Took Over the Democratic Party*. President Obama, according to Von Bulow, 'has taken over health care – nationalize health care, where every person in the U.S. will be controlled by the government. That's FASCISM!!!' (2010: 12).

In *The Birth of Biopolitics*, Foucault describes the neoliberal stance toward the state as one of repulsion or phobia. This is the fear of the 'unlimited growth of the state, its omnipotence, its bureaucratic development, the state with the seeds of fascism it contains, the state's inherent violence beneath its social welfare paternalism' (2008: 186–187). Like Von Bulow, outspoken radio and television personality Glenn Beck articulates this fear of the 'seeds of fascism' by asserting that Obamacare 'is going to come out the other side dictorial [*sic*] – it's going to come out a fascist state' (2009).[5] Even if the proposed intervention is beneficial or intends to be beneficial, critics of State-based welfare programmes believe that the 'seeds of fascism' are present. Despite the fringe status of Beck and Von Bulow, they present a neoliberal phobia of the State that maintains that government programmes such as universal healthcare will lead at best to bureaucratic inefficiency, and at worst to totalitarianism.

The fear of collectivist social policy, particularly under the banner of the 'welfare state', predates the recent paranoia exhibited by pundits like Beck in America. In 1944 Hayek denounced collective social planning in the *Road to Serfdom*. Although Hayek's thoughts on the size and role of the state can be at times ambiguous, he argued that centralized state planning required the rule of the few over the many, which even if initially benevolent, will ultimately lead to unjustified restrictions on individual freedoms (2001). Similarly, in 1949 President Hoover was quoted in *Life* magazine stating 'the slogan of a "welfare state" has emerged as a disguise for the totalitarian state' (Garland 2014, 333 n. 14). These critiques of State intervention in the economic and social realms can be understood as a reaction against the purported ideals of the 'welfare state', or what Thatcher referred to as the 'collective society' of the post-war period (Butt 1981). Rather than central planning and bureaucracy, neoliberal governmentality contends that the market of free participation by individuals is the best form of welfare and social organization.

### *Manipulating* homo œconomicus

The idea of a small and limited government not only reflects an economic and political theory, but also expresses an anthropological perspective. That is, a perspective on the human condition and what it is to be a human being. According to Howell and Ingham, neoliberalism was more than an economic vision: it was 'also an affective and moral philosophy that, in embracing voluntarism and individualism, required all citizens to do something for and about themselves' (Howell and Ingham 2001: 330). This self-interested and entrepreneurial individual is what economic and political theorists refer to as *homo œconomicus* (Wilson and Dixon 2012).

Foucault describes *homo œconomicus* – his central point of reference in *The Birth of Biopolitics* lectures (Lemke 2001: 200) – as 'someone who is eminently governable' (2008: 270). The governability of the neoliberal subject is the result of the 'certain form of freedom' operating in neoliberal theory (Foucault 2007: 353). In contrast to the freedom of classical liberalism, in which *homo œconomicus* has an intrinsic and natural freedom that limits government (Lemke 2001: 200), the neoliberal subject has an entrepreneurial freedom of choice that enables governance. Foucault articulates *homo œconomicus* as an entrepreneur of the self by drawing on Gary Becker's human capital theory (Foucault 2008: 226, Lemke 2001: 199). Human capital theory transforms human labour from something exchanged for a wage or embedded in the production of a commodity into a 'subjective *choice*' made by the labourer (Dilts 2011: 135). Rather than exchange labour for a wage, the subject invests their human capital in order to receive an income. Foucault's *homo œconomicus* has an entrepreneurial relation to himself and their activity – 'being for himself his own capital, being for himself his own producer, being for himself the source of [his] earnings' (Foucault 2008: 226). By re-conceiving the subject as an entrepreneur of himself, all choices and activities of life, not just labour, are transformed into investments and incomes that 'may or may not improve human capital' (Foucault 2008: 230). In rethinking the subject as an entrepreneur who is free to choose to invest in herself, neoliberal *homo œconomicus* becomes governable through systematic 'modifications in the variables of the environment' (Foucault 2008: 270, Gordon 1991: 43).

The idea of manipulating the environment of *homo œconomicus* has been popularly put forward by Richard Thaler and Cass Sunstein in their book, *Nudge: Improving Decisions About Health, Wealth and Happiness* (2009).[6] Thaler and Sunstein's approach, which they term libertarian-paternalism, encourages governments and private institutions to guide individuals toward choices that will improve their lives. In this way, libertarian-paternalism is an expression of neoliberal governmentality that has the objective of security (paternalism), achieved through individual freedom (libertarian).

For Thaler and Sunstein, the paternalistic logic operates through 'choice architecture'. Choice architecture is the systemic structuring of choices to enable individuals to make the 'right' or 'best' choice with the least amount of resistance. An example they use is ordering food in a school cafeteria – if fruits and salads are placed toward the front they are more likely to be chosen than if placed at the back. Thaler and Sunstein argue that 'small and apparently insignificant details can have major impacts on people's behaviour' and that 'the power of these small details comes from focusing the attention of users in a particular direction'

(2009: 3–4). The goal of the choice architect is to manipulate the environment or nudge 'people's behaviour in a predictable way without forbidding any options or significantly changing their economic incentives' (2009: 6). Thus the freedom of the individual is both preserved and used to govern the individual towards choices and behaviours that are considered by the architect to be in the best interests of the individual and population. Thaler and Sunstein's argument for libertarian paternalism operates with the neoliberal logics of individual freedom, choice and market mechanisms.

In addition to topping best-seller lists, *Nudge* has influenced the policy frameworks of the Obama Administration in the US and the Cameron Government in the UK (McSmith 2010), including public health policy and anti-obesity campaigns (Hickman 2010, Secretary of State for Health 2010a, 2010b, Thaler and Sunstein 2010). Andrew Lansley, former Secretary of State for Health in the Cameron Government, drew on Thaler and Sunstein's ideas when re-modelling UK public health policy. Lansley wanted to promote 'the Government's core values of freedom, fairness and responsibility by strengthening self-esteem, confidence and personal responsibility' (Secretary of State for Health 2010a: 6). To achieve this he emphasized the 'freedom ensuring' approach of nudging in contrast to the 'freedom depriving' strategies of the welfare state or what is pejoratively called the nanny state. For Lansley, the role of government or public health authorities is not to tell people what to do, but to nudge 'individuals in the right direction. Encouraging positive choices. Not lecturing or nannying. But making people feel empowered' (Secretary of State for Health 2010b). By drawing on the logics of neoliberal governmentality, or nudging, which uses individual freedom to secure the population, Lansley and the Cameron Government intend to fulfil their 'commitment to protecting the population from serious health threats' by shifting responsibility and choice on to the individual (Secretary of State for Health 2010a: 4).

The manipulation of the environment is not a direct intervention, but neither is it the deep structural change that social determinants of health researchers advocate (Marmot 2005, Cheng 2012, Brownell et al. 2010). Rather, neoliberal governmentality attempts to shape education, health and the self through financial calculations and judgements that operate 'indirectly, through contracts, targets, performance measures monitoring and audit' (Rose 1999: 151). As market logics permeate everyday life, the individual is presented with new and entrepreneurial 'way[s] to develop the self... [that] offer a highly elastic mode of self-mastery that channels doubt over uncertain identity into fruitful activity' (Martin 2002: 9). Sam Binkley writes that neoliberal governmentality employs mechanisms, such as lifestyle, to govern subjects as 'market agents, encouraged to cultivate themselves as autonomous self-interested individuals, and to view their resources and aptitudes as human capital for investment and return' (2009: 62). This process reconceives health or education as resources that the individual can invest in, promote and secure through choices, but for which the individual is ultimately responsible.[7]

## Obesity and three biopolitical tensions: freedom and security, individual and population, present and future

The influence of neoliberal ideas produces the political and economic conditions for the enabling network of lifestyle to emerge and link individual choice to population security. The entrepreneurial rhetoric of present investment and future returns is a strong theme in the *Let's Move* campaign initiated by the US First Lady, Michelle Obama. In launching the campaign, Obama appealed to everyone to get involved and secure the future health of the nation – 'we'll need to make some modest, but critical, investments in the short-run... but we know that they'll pay for themselves – likely many times over – in the long-run' (Office of the First Lady 2010). The idea of investing in one's self and children is a key tool 'to govern behaviour at all levels of the social organism' (McNay 2009: 58). Through the rhetoric of choice, responsibility and self-enterprise, neoliberal governance manages threats and secures individuals and the population. In remapping individual freedoms as responsible self-management, 'individual autonomy is not an obstacle or limit to social control but one of its central technologies' (McNay 2009: 63). Thus, freedom – rather than a ban or law – is mobilized to encourage individuals to transform the self into an enterprise directed toward the biopolitical end of a healthy lifestyle.

Guiding choice in relation to chronic disease has become a central strategy of neoliberal health governance. Michael Dillon and Luis Lobo-Guerrero note that '[w]hereas infectious disease invited drug cures, chronic illnesses invite pre-emption and prevention together with medical campaigns aimed at changing life-styles' (Dillon and Lobo-Guerrero 2009: 285). Michelle Obama echoes this point, when she says that obesity 'isn't like a disease where we're still waiting for the cure to be discovered – we know what the cure for this is... We have everything we need, right now, to help our kids lead healthy lives' (Office of the First Lady 2010). According to Obama and *Let's Move*, individuals need to choose to exercise and eat healthier food more often in order to solve this 'problem of such magnitude' (Office of the First Lady 2010). Thus in the context of the obesity epidemic, the use of individual choice is central to securing the population from catastrophe.

The neoliberal rhetoric of politicians and public health campaigns characterize everyday individual choices and activities as causing large-scale epidemics, environmental disaster or threats to national security. Activities such as playing computer games, microwave dinners, sugary drinks, jogging, infant feeding, alcohol consumption, prenatal yoga, mode of transportation, or outdoor-dress are not only indicators of individual health or disease risk, but present a threat to population security (Holland et al. 2011, Lupton 2004, Gard and Wright 2005). These activities become political, and those engaging in them become legitimate targets for further intervention. The lifestyle network positions the individual as free and uniquely responsible 'for taking pre-emptive countermeasures to ensure *a future that is continuous with the past*' (Diprose et al. 2008: 270). Rather than allowing present and unpredictable daily activities and bodies of individuals to morph into future threats to populations, individuals must be in a constant state of readiness, and be prepared to adopt pre-emptive measures. The daily activities and choices of individuals are no longer benign and inconspicuous, but are made visible as potential threats to the population requiring governance and direction in order to mitigate the possibility of catastrophe at the level of population.

## Dynamics of obesity

Over the past two decades obesity has transformed from an individual problem with predictable and isolated consequences into a population wide calamity with unpredictable consequences that threatens social and economic stability (AAP 2010, Andrews 2011, Associated Press 2010, Bartlett 2008, *Sydney Morning Herald* 2010). Obesity is framed as a unique phenomenon, unlike previous public health crises, and as such requiring new techniques of governance. Unlike smoking, unprotected sex, or intravenous drug-use, the need to eat is universal and therefore everyone is theoretically at risk of becoming obese. Characterized as an imbalance between energy in (food and drink) and energy out (exercise) that leads to a BMI of >30, the ontology of obesity is premised on activities fundamental to the maintenance of human life. Obesity is therefore a condition to which everyone is supposedly susceptible. Further, obesity is dynamic. The equilibrium between energy in and energy out is always at risk of imbalance. As such, individuals are never 'safe' from obesity by virtue of avoiding a particular behaviour or activity, but always exist on a spectrum, either becoming obese or becoming 'normal'. This makes obesity uniquely difficult to govern. To govern the dynamic and ubiquitous threat of obesity, health authorities and governments have introduced a range of strategies and campaigns that target the entire population in order to modify and guide individual choice towards health promoting activities.

Although obesity at the level of the individual can be meticulously measured, calculated and controlled, at the population level it is portrayed as an unpredictable threat emanating from the accumulation of these individual choices. The dynamic between individual choice and population effects therefore requires mechanisms of governance to secure the unknown, incalculable and unpredictable. These logics can be seen in areas such as the control of the spread of infectious disease or counter-terrorism measures, which involve the management of dynamic and catastrophic risks. Likewise, the intensification of focus on obesity has led to an increase in techniques, methods and strategies to pre-empt and predict (Diprose et al. 2008).

Three tensions characterize the biopolitical rationality that attempts to secure the population by pre-empting future threats in contemporary governance: present–future, freedom–security, and individual–population. The tension between the present and the future is depicted as continuous and volatile. A seemingly benign event, act or practice in the present is perceived as containing the potential for catastrophe in the future. The tension between freedoms and security provides the condition and objective of the political rationality. Within this rationality, strategies of security should not conflict with freedoms but use individual freedom to secure the population. Finally, the tension between the individual and the population is the culmination of the previous relationships: the individual in the present is required to use their freedom in order to secure the population of the future. Through these tensions, the individual is drawn from an undifferentiated population and positioned as a subject to be targeted and governed. I provide a brief sketch of these tensions.

### Present–future

Attempts to order the present as a means of warding off harmful events in the future are not unique to the biopolitics of security. What is unique, however, is the multiplication of knowledges and techniques that seek to close the temporal gap between the present and future, and to wholly control the relationship between the two – so that a controlled present will continue into a controlled future. The rationalities of security tend to produce subjects attentive to present choices out of a concern for future consequences. The extension of the present into the future encourages individuals to think of the present from the perspective of the future. The imminent yet incalculable nature of the threat requires present vigilance in order to secure the future. In emphasizing the responsibility of individuals for the future consequences of present choices and actions, the role of freedom and its relation to security becomes central. According to François Ewald, the notion of responsibility is bound to the advent of liberalism, and involves producing subjects 'aware of the future, and preventing them from living solely in the present' (2002: 274–275). For example, life insurance and retirement planning schemes encourage individuals to ensure the economic future of their family remains secure whether or not they are employed, unemployed or deceased. As individuals are directed to be aware of the future in present activity, the consequences of an action mushroom, and responsibility is broadened to include the unintended and unforeseeable.

### Freedom–security

The second tension in the biopolitics of security is that between the freedom of the individual and the security of the population. These are both recoded in the lifestyle network, where freedom is used to shape and direct choices towards security. Foucault argues that the 'fundamental objective of governmentality will be mechanisms of security', and 'a condition of governing well is that freedom, or certain forms of freedom, are really respected' (2007: 353). The goal of security is thus achieved not by limitation or prohibition of individual movement, activity or choice, but through avenues of freedom. Freedom and security dovetail into one another such that 'freedom is nothing else but the correlative of the deployment of apparatuses of security… [which] cannot operate well except on condition that it is given freedom' (2007: 48). Thus the free choice is the choice that promotes health and security.

*Individual–population*

The fusion of the present with the future, and of freedom with security, draws into focus the relationship between the individual and the population. Governance through freedom to achieve security of the biological life of the population produces and targets subjectivities, through techniques of discipline, regulation and normalization. The governmental concern with population security and incalculable risk produces a corresponding turn towards the individual as the vector of risk and agent of security (Wilkinson 2010: 55). Rather than go against neoliberal ideals and intervene in the activities of corporations, professional organizations or the scientific community, governments instead mobilize strategies that modify the environment of 'individuals and particular social groups whose behaviour comes to be viewed as a risk to general physical security' (Diprose et al. 2008: 271).

The tensions between individual and population, freedom and security, and present and future that play out in the rhetoric and governance of obesity introduce new strategies to secure the effects of individual choice and behaviour. Stefan Elbe writes that 'the construal of obesity as a deeper threat to the population also opens yet another domain of social life to more intensive forms of regulation' (2010: 153). The intensity of regulation is not simply disciplinary, but concentrates on conducting the conduct of individuals from a distance (Rose, O'Malley, and Valverde 2006). In this regard, strategies such as social marketing are increasingly deployed.

## Measure up and the strategy of social marketing

The characterization of obesity as a security threat emanating from individual choice and commercial practices poses a problem for neoliberal governmentality. Public health researchers suggest policies such as 'fat taxes', or bans on sugar-sweetened beverages (Brownell et al. 2009, Caraher and Cowburn 2005, Jacobson and Brownell 2000, Thow et al. 2011). Although these suggestions are based on evidence supporting their efficacy, they are excluded from the neoliberal arrangement due to their potential impact on the freedom of markets and individuals (Marlow, Trinko 2013, Snowdon 2013). Rather than intervening directly via industry regulations or restricting choice, governments in Australia, the UK and the US deploy social marketing strategies to shape individual choice and behaviour at a distance (Blumenthal-Barby and Burroughs 2012, Ménard 2010, Rayner and Lang 2011, Minister for Health and Ageing 2009).

The birth of social marketing is often traced to Gerhardt Wiebe. Around the same time that Hayek and Hoover expressed concern over collective welfare programmes, Wiebe asked the American Psychology Association: 'Why can't you sell brotherhood and rational thinking like you sell soap?' (Wiebe 1951). This question prompted researchers and policy-makers to explore the potential for commercial marketing techniques to motivate social change, in policy areas such as juvenile delinquency, neighbourhood engagement and cholesterol awareness

(Kotler and Zaltman 1971, Laczniak, Lusch, and Murphy 1979, Lefebvre and June 1988). Since Wiebe's initial question, the definition and practice of social marketing have evolved (Smith 2000, Truong 2014). Rather than simply advertising social messages, social marketing has become a coordinated strategy that uses components of commercial marketing, such as: a consumer orientation; an emphasis on voluntary exchanges between providers and consumers; audience research and segmentation strategies; the use of formative research in material design; focus groups to pretest materials; diverse distribution channels; use of the marketing mix (product, price, place, and promotion); and monitoring and follow-up analyses (Grier and Bryant 2005, Lefebvre and June 1988).

The coordinated approach to social marketing can be seen in the Australian Government's *Measure Up* campaign.[8] The *Measure Up* campaign was launched nationally in October 2008, to raise awareness and reduce incidence of weight-related diseases among the Australian population. Using television, radio, internet, print and outdoor media, the campaign encouraged Australians to monitor their diets and exercise habits in order to achieve a waistline measurement associated with 'normal' health – a waistline between 66–80 cm for women, and 76–94 cm for men.

While the campaign was overseen by the Australian Department of Health (DoH), the complexity of conducting social marketing led government health authorities to collaborate with commercial research companies to design and implement campaign materials (Lupton 2014). Market research company GFK Bluemoon was contracted by the DoH to conduct focus groups to determine the attitudes and beliefs of the target group in relation to lifestyle change. The stated aim of the research 'was to explore whether the threat of chronic disease can be leveraged effectively in communications to stimulate behaviour change' (GFKBM 2007: 4). According to GFK Bluemoon, the campaign needs to make individuals aware of their 'susceptibility' to the 'threat of chronic disease' by 'conveying the severity of these conditions that can result from inaction' (GFKBM 2007: 7).

While the brief may have been to communicate lifestyle messages to all Australians, GFK Bluemoon found through focus groups that not all people would respond to a lifestyle social marketing campaign in the same manner. Using the social marketing strategy of audience segmentation, GFK Bluemoon divided the research participants into six subgroups: 'endeavourers', 'balance attainers', 'defiant resisters', 'quiet fatalists', 'apathetic postponers' and 'help seekers' (GFKBM 2007 39). The 'endeavourers' and 'balance attainers' were people who recognized the importance of a 'healthy weight' and the lifestyle changes that were needed, why those changes are important, and how they can be achieved. As such, people in these two subgroups were considered to be '"at lower risk" of developing lifestyle related chronic diseases' (GFKBM 2007 5). In contrast, the 'defiant resisters', 'quiet fatalists', 'apathetic postponers' and 'help seekers' were people who had a low appreciation for why lifestyle change is needed and/or how those changes can be made. The aim of the campaign was 'to migrate as many people as possible to "endeavourer" and "balance attainer" segments' (GFKBM 2007: 7). The 'apathetic postponers' and 'help seekers' were identified as the segments most likely to migrate into more desirable segments. GFK Bluemoon therefore recommended that the campaign should focus on those subgroups, as these 'people already had some appreciation of "what" change is required and therefore found this approach highly credible and motivating' (GFKBM 2007: 6).

The 'defiant resisters' and 'quiet fatalists', however, were considered to be unlikely to respond positively to a social marketing campaign. Although these subgroups had a higher percentage of people affected by chronic disease, GFK Bluemoon acknowledges that socio-economics, structural factors and cultural background are significant barriers to making lifestyle changes.

> There appears to be a strong correlation between attitudes to change and socio-economic factors. People from socially disadvantaged groups, including NESB [non-English speaking background] and Aboriginal and Torres Strait Islander communities, were over-represented in the 'Defiant Resister' and 'Quiet Fatalist' segments.
>
> (GFKBM 2007: 69)

While a social marketing campaign is 'likely to have a more immediate and obvious influence' on the lives of people with social and economic privilege, GFK Bluemoon acknowledges that 'those from socially disadvantaged groups are faced with serious structural and environmental barriers to changing their behaviour, irrespective of their attitudes to change' (GFKBM 2007: 70). Nevertheless, the campaign was not modified to respond to these barriers that affected those supposedly at greatest risk. Instead, those considered 'too resistant' were excluded from the focus of the campaign altogether. Rather than addressing the structural and environmental barriers of those most affected by chronic disease (which would require large-scale government intervention) the social marketing campaign focused on the segments of the population with the social and economic resources able to make the proposed lifestyle changes.

Based on this research the campaign had four objectives:

1 To increase awareness of the link between chronic disease and lifestyle risk factors (poor nutrition, physical inactivity, unhealthy weight);
2 To raise appreciation of why lifestyle change should be an urgent priority;
3 To generate more positive attitudes towards achieving recommended changes in healthy eating, physical activity and healthy weight;
4 To generate confidence in achieving the desired changes and appreciation of the significant benefits of achieving these changes (Australian Better Health Initiative 2007a).

The centrepiece of the *Measure Up* campaign was a 60-second television commercial that was aired throughout Australia.[9] The commercial uses computer-generated-imagery (CGI) to vividly show the transformation of a 20-something man with a waist measurement of 84 cm, who ages and slowly gains weight as he walks along a large orange tape-measure toward the camera. At the 92 cm mark he casually opens the commercial with, 'You know how it is [*pause*] you settle down [*pause*] put on a few kilos [*pause*] but I'm not worried.' A calm female voiceover provides authoritative and statistical information, stating, 'One in two Australian adults is overweight.' At the 96 cm mark the man states, 'Life gets busier, you let yourself go a bit [*pause*] I'm not worried.' The voiceover responds with, 'Unhealthy eating and drinking and not enough physical activity can seriously affect your health.'

At the 102 cm mark the man has aged considerably, looks uncomfortable as he walks and is unable to play with his young daughter, who appears disturbed to see her father struggling and out of breath. The voiceover informs the viewer that, 'For most people, waistlines of over 94 cm for men and 80 cm for women increase the risk of some cancers, heart disease and type 2 diabetes.' At this point the man, out of breath with his hands on his knees, lethargically states, 'When I realized it was affecting my health [*pause*] well, yeah, I got worried.' Concluding the commercial, the voiceover informs the viewer, 'The more you gain, the more you have to lose' (Australian Better Health Initiative 2007b).

This commercial was designed to shock and produce an emotional response. According to GFK Bluemoon's research, 'Shocking imagery, in the style of recent anti-smoking campaigning, seemed to have a great deal of impact and was surprisingly acceptable to the majority of people' (GFKBM 2007: 6). The message of the television commercial was reinforced by radio advertisements, posters on bus shelters and digital advertisements on prominent news, entertainment and social networking websites. The motif of measuring in general, and tape measures in particular, was used to establish a network of choice, risk and chronic disease. According to the initial media release, 'it's not just about measuring your waist – what *Measure Up* is really measuring is risk' (Roxon 2008). The campaign does not conceive of risk as a threshold (one is either at risk or not), but as a spectrum, on which one always has some level of risk. The risk level is determined through the measurement of the waist circumference (Australian Better Health Initiative 2007b: 2).

The tone of the campaign was designed by GFK Bluemoon to provoke anxiety to 'leverage' concerns about weight-related chronic disease to encourage change in behaviour (GFKBM 2007). This was achieved through the CGI commercial and tape measure. The campaign-issued tape measure provides visual and text based descriptions of risk. Inscribed on the tape measure is the statement: 'If your waist measures more than 80 cm [94 cm for men], you may be increasing your risk of chronic diseases such as some cancers, heart disease and type 2 diabetes. Greatly increased risk = 88 cm [102 cm for men] or more' (Australian Better Health Initiative 2007a). Accompanying the text is the 'healthy range' represented by a green band that fades in from white at 66 cm (76 cm for men) and fades into orange at 80 cm (94 cm for men), which then fades at 88 cm (102 cm for men) into the red 'high risk' range. Anxiety was also used in targeted settings to motivate change. Supermarket shopping trolleys carried advertisements asking shoppers, 'How does your trolley measure up?' and reminding them that 'Poor food choices and an inactive lifestyle increase your risk of developing chronic diseases' (Australian Better Health Initiative 2007a).

Importantly, in this campaign the goal of a 'healthy lifestyle' is not imposed on the individual, but is marketed as something that is achievable and desirable for the individual to adopt. The campaign also used positive messages to entice individuals to take responsibility for their health and transform negative habits into a positive lifestyle. Testimonials, recipes, exercise regimes and expert advice are offered via the campaign website and other materials to generate confidence and appreciation of the benefits of changing, as per the fourth campaign objective.

The positive and enticing messaging was further reinforced in a follow-up campaign, *Swap It, Don't Stop It*. Launched as a 'new phase of *Measure Up*', the *Swap It, Don't Stop It* campaign was designed to show people 'how they can take steps to help reduce their waist measurement and improve overall health and wellbeing' (Department of Health and Ageing 2011). The *Swap It, Don't Stop It* phase uses a more playful and light-hearted tone. Instead of an ageing, obese and lethargic man, the campaign is fronted by an animated blue balloon character called Eric. According to the Health Minister,

> Eric will urge Australians to make some simple lifestyle changes to become healthier – for example, to swap big for small (portion control); swap often for sometimes (occasional treats); swap fried for fresh (nutritional quality); swap sitting for moving (physical activity); and swap watching for playing (physical activity).
>
> (Department of Health and Ageing 2011)

The concept and practice of a 'swap' encourages the individual to adopt an entrepreneurial attitude in negotiating choice in everyday life. Through encouraging the individual to choose freely – and to continue consuming – *Swap It, Don't Stop It* seeks to 'combat' the threat of obesity (Department of Health and Ageing 2011). The campaign aims to make visible incidental choices and to encourage modification – as 'those little decisions made in the supermarket aisles, in the kitchen or when playing with the kids, can make a real difference' (Roxon 2011). By adopting an attitude of swapping, exchange or bartering in everyday life – or as the campaign slogan suggests, 'to become a swapper' – the individual fashions a lifestyle that promotes individual health and secures the health of the population from obesity.

The *Measure Up* social marketing campaign, and its *Swap It, Don't Stop It* phase, have been critically examined for their efficacy, ethics and government collaboration with commercial research groups such as GFK Bluemoon (Lupton 2014, Carter et al. 2011). However, there has been little analysis of these campaigns as strategies of neoliberal governmentality (Crawshaw 2012). Public health social marketing campaigns and discourses position individual choice as a central site requiring governance. Encouraging individuals to responsibly self-govern via choice reinforces the neoliberal political and economic ideas of freedom and individuality. As such, the three biopolitical tensions described above – freedom and security, individual and population, present and future – all operate through the *Measure Up* social marketing campaign.

The *Measure Up* campaign does not restrain the individual from acting freely. Rather, free choice is reconfigured to entail choosing responsibly and securely. The CGI, print-advertisements and text of the campaign entice the individual to freely make choices and to modify behaviour in a manner that promotes health and security for the population. The CGI and population statistics combine to modify the environment of living beings to secure the population via the individual. However, this modification is not direct: it does not involve either removing or forcing choice. Choice remains free, yet the campaign seeks to make certain choices within the environment visible as irresponsible in relation to the population. The rhetoric and visuals of the campaign characterize individual choice as accumulative, and which, if irresponsible, will lead to obesity that has an adverse effect on the population. The responsible subject is encouraged to choose in order to promote their own health and to secure the health of the population.

In addition to the freedom and security, and individual and population, relations, the neoliberal governmentality of obesity also modifies the relation between present and future. Individuals are not only responsible for the immediate present, but also for the future consequences of present choices and actions. The *Measure Up* campaign blurs the relation between present and future through the use of CGI to project an obese future emanating from an irresponsible present. Present and future are conflated in order to govern the ontology of obesity as a chronic and dynamic condition that manifests over time, and is always in a state of 'becoming'. To govern the temporal, dynamic and ubiquitous threat of obesity, social marketing strategies like *Measure Up* position individuals as free yet always responsible. The CGI of *Measure Up* extends the present into the future, encouraging individuals to think of the present from the perspective of the future and to take 'responsibility for the future inherent in present acts' (Diprose et al. 2008: 270). If the individual is not vigilant in the present, then weight, disease and anxiety may unexpectedly emerge in the future. The uncertain possibility of becoming obese is reinforced by health authorities declaring that the threat of obesity and disease cannot be eliminated, but that 'healthy eating [with] regular physical activity helps you maintain good health, reduces your risk of chronic diseases such as some cancers, heart disease and type 2 diabetes, and can help prevent obesity' (Australian Better Health Initiative 2007b: 6). Thus, unlike calls to quit smoking and thereby abolishing the risk of lung disease, anti-obesity rhetoric calls for continual vigilance and self-monitoring. Without individual vigilance, the not-worried 20-something becomes a worried and obese 40-something.

Neoliberal strategies of governance recode freedom and security, transforming the relationship between individuals and populations. In this context, lifestyle becomes a biopolitical mechanism that makes the incidental choices of individuals visible as objects requiring responsible self-governance. Strategies such as social marketing, and techniques such as CGI and tape measures, create the conditions under which the gap between present choices and future outcomes is removed, and it is possible to imagine that a controlled present will continue into a controlled future. The neoliberal rationality operating in health policy and health promotion campaigns contributes significant lines of power and knowledge to the lifestyle network. In doing so, it mobilizes a specific notion of freedom to enable the governance of individuals and populations. The objective of security and the condition of freedom produce campaigns such as *Measure Up* and *Swap It, Don't Stop It*, which seek to influence and manipulate the choice environment, such that the free choice aligns with the responsible choice that secures the population. While this chapter has focused on the technologies of power operating in neoliberal governmentality, later chapters will emphasize the importance of technologies of the self. Prior to this, however, the next chapter explores a further feature of the lifestyle *dispositif*, the lines of knowledge that establish obesity as an 'urgent need' requiring governance.

## Notes

1   These figures vary depending on the source. For Australia see (Australia Parliament House of Representatives Standing Committee on Health and Ageing 2009, Colagiuri et al. 2010); for the US see (Tsai, Williamson, and Glick 2011); and the UK see (Public Health England 2015).

2   For the UK context see (Secretary of State for Health 2010a) and for the US see (White House Task Force on Childhood Obesity 2010).

3   It is not necessary to provide a detailed account of liberalism or Ordoliberalism here, as my interest is in the impact of Neoliberalism on lifestyle, rather than the influence of Ordoliberalism/liberalism on Neoliberalism. Therefore, I focus on Neoliberalism and take the background of liberalism and Ordoliberalism as given. For a discussion of the relationship between Ordoliberalism and Neoliberalism in Foucault see (Lazzarato 2009) and (Lemke 2001).

4   Milton and Rose Friedman use children and madmen as examples of irresponsible choosers who can justifiably have their freedom restricted by the state (Friedman and Friedman 1980: 32).

5   Lars Thorup Larsen notes the influence of neoliberal state-phobia on health policy in commenting, 'health programs are clearly written against the backdrop of state-phobic critiques that state-based health policies have undermined the responsibility of both individuals and communities'. (Larsen 2011: 222).

6   Thaler and Sunstein reject the term *homo oeconomicus*, arguing that human beings are not capable of rational choice in the manner conceived by economists. This rejection cannot be adequately addressed here; however, I contend that they are rejecting the liberal *homo oeconomicus* and although they do not use the term they are affirming neoliberal *homo oeconomicus*, who responds to modifications in the choice environment.

7   Slavoj Žižek sardonically summarizes the situation of uncertainty under neoliberalism: 'you have to change job every year, relying on short-term contracts instead of long-term stable appointment. Why not see it as the liberation from the constraints of a fixed job, as the chance to reinvent yourself again and again, to become aware of and realize hidden potentials of you personality? You can no longer rely on the standard health insurance and retirement plan, so that you have to opt for additional coverage for which you have to pay. Why not perceive it as an additional opportunity to choose: either better life now or long-term security?' (2001: 116).

8   For a critical analysis of the UK social marketing campaign, *Change4Life* see (Coleman 2013).

9   This commercial can be seen on YouTube by searching 'How Do You Measure Up?'

# References

AAP. 2010. "Obesity, climate change are 'great threats': doctors." *Herald Sun*, March 14.

Alejandro Leal, Pablo. 2007. "Participation: the ascendancy of a buzzword in the neo-liberal era." *Development in Practice* 17 (4–5):539–548. doi: 10.1080/09614520701469518.

Andrews, Winnie. 2011. "'Tsunami' of obesity worldwide: study." *Sydney Morning Herald*, February 4.

Associated Press. 2010. "Obesity bigger threat than terrorism?" *CBS News*, July 17.

Australia Parliament House of Representatives Standing Committee on Health and Ageing. 2009. *Weighing it Up: Obesity in Australia*. edited by House of Representatives Standing Committee. Canberra: Printing and Publishing Office House of Representatives.

Australian Better Health Initiative. 2007a. *Measure Up*. Australian Government 2007 [cited November 17, 2009]. Available from www.measureup.gov.au.

Australian Better Health Initiative. 2007b. *Time to Take Some Healthy Measures? How Do You Measure Up?* Australian Government 2007 [cited August 15 2009]. Available from http://www.australia.gov.au/MeasureUp.

Bartlett, Lawrence. 2008. "Obesity more dangerous than terrorism: experts." *The Age*, February 25.

Beck, Glenn. 2009. "Beck: Healthcare 'system is going to come out the other side dictatorial – it's going to come out a fascist state'." *Media Matters for America* 2009 [cited 6 May 2015]. Available from http://mediamatters.org/video/2009/07/27/beck-healthcare-system-is-going-to-come-out-the/152494.

Berlin, Isaiah. 2002. "Two concepts of liberty." In *Liberty*, edited by Henry Hardy. New York: Oxford University Press.

Biancardi, Fabian. 2003. *Democracy and the Global System: A Contribution to the Critique of Liberal Internationalism*. Basingstoke: Palgrave Macmillan.

Biltekoff, Charlotte 2007. "The terror within: obesity in post 9/11 U.S. life." *American Studies* 48 (3). https://journals.ku.edu/index.php/amerstud/article/view/3132/3911

Binkley, Sam. 2009. "The work of neoliberal governmentality: temporality and ethical substance in the tale of two dads." *Foucault Studies* 6:60–78.

Blumenthal-Barby, J.S., and Hadley Burroughs. 2012. "Seeking better health care outcomes: the ethics of using the 'nudge'." *American Journal of Bioethics* 12 (2):1–10. doi: 10.1080/15265161.2011.634481.

Boseley, Sarah. 2015. "David Cameron's plans for obese benefit claimants questionable, says the Lancet." *The Guardian*, 19 March.

Brownell, Kelly D., Thomas Farley, Walter C. Willett, Barry M. Popkin, Frank J. Chaloupka, Joseph W. Thompson, and David S. Ludwig. 2009. "The public health and economic benefits of taxing sugar-sweetened beverages." *New England Journal of Medicine* 361 (16):1599–1605. doi: doi:10.1056/NEJMhpr0905723.

Brownell, Kelly D., Rogan Kersh, David S. Ludwig, Robert C. Post, Rebecca M. Puhl, Marlene B. Schwartz, and Walter C. Willett. 2010. "Personal responsibility and obesity: a constructive approach to a controversial issue." *Health Affairs* 29 (3):379–387. doi: 10.1377/hlthaff.2009.0739.

Butt, Ronald. 1981. "Mrs Thatcher: the first two years." *Sunday Times*, 3 May.

Cahill, Damien. 2014. *The End of Laissez-Faire?: On the Durability of Embedded Neoliberalism*. Cheltenham: Edward Elgar.

Caraher, Martin, and Gill Cowburn. 2005. "Taxing food: implications for public health nutrition." *Public Health Nutrition* 8 (08):1242–1249.

Carmona, Richard H. 2003. *Remarks to the American Medical Association's National Advocacy Conference.* edited by U.S. Department of Health and Human Services. Washington DC: U.S. Department of Health and Human Services.

Carter, Stacy M., Lucie Rychetnik, Beverley Lloyd, Ian H. Kerridge, Louise Baur, Adrian Bauman, Claire Hooker, and Avigdor Zask. 2011. "Evidence, ethics, and values: a framework for health promotion." *American Journal of Public Health* 101 (3):465–472. doi: 10.2105/ajph.2010.195545.

Cheng, Jennifer K. 2012. "Confronting the social determinants of health—obesity, neglect, and inequity." *New England Journal of Medicine* 367 (21):1976–1977.

Colagiuri, Stephen, Crystal M. Lee, Ruth Colagiuri, Dianna Magliano, Jonathan E. Shaw, Paul Z. Zimmet, and Ian D. Caterson. 2010. "The cost of overweight and obesity in Australia." *Medical Journal of Australia* 192 (5):260–264.

Coleman, Rebecca. 2013. *Transforming Images: Screens, affect, futures.* New York: Routledge.

Crawshaw, Paul. 2012. "Governing at a distance: social marketing and the (bio)politics of responsibility." *Social Science & Medicine* 75 (1):200–207. doi: http://dx.doi.org/10.1016/j.socscimed.2012.02.040.

Deleuze, Gilles. 1992. "What is a dispositif?" In *Michel Foucault, Philosopher*, edited by Timothy J. Armstrong. New York: Routledge.

Department of Health and Ageing. 2011. *Campaign Fact Sheet – Swap it, Don't Stop it.* Australian Government 2011a [cited August 31 2011]. Available from http://swapit.gov.au/resources/downloads/campaign-fact-sheet.

Devine, Pat. 2007. "The 1970s and after: the political economy of inflation and the crisis of social democracy " In *Reading Karl Polanyi for the Twenty-First Century*, edited by Ayse Bugra and Kaan Agartan. New York: Palgrave Macmillan.

Dillon, Michael, and Luis Lobo-Guerrero. 2009. "The biopolitical imaginary of species-being." *Theory, Culture & Society* 26 (1):1–23. doi: 10.1177/0263276408099009.

Dilts, Andrew. 2011. "From 'entrepreneur of the self' to 'care of the self': neo-liberal governmentality and Foucault's ethics." *Foucault Studies* 12:130–146.

Diprose, Rosalyn, Niamh Stephenson, Catherine Mills, Kane Race, and Gay Hawkins. 2008. "Governing the future: the paradigm of prudence in political technologies of risk management." *Security Dialogue* 39 (2–3):267–288.

Elbe, Stefan. 2010. *Security and Global Health.* Cambridge: Polity.

Ewald, François. 2002. "The return of Descartes's malicious demon: an outline of a philosophy of precaution." In *Embracing Risk: The Changing Culture of Insurance and Responsibility*, edited by Tom Baker and Jonathan Simon. Chicago, IL: University of Chicago Press.

Foucault, Michel. 2000. "The risks of security." In *Power: Essential Works of Foucault 1954–1984*, edited by James D. Faubion. Harmondsworth: Penguin Books.

Foucault, Michel. 2007. *Security, Territory, Population: Lectures at the Collège de France 1977–78.* Translated by Graham Burchell. Edited by Arnold I. Davidson. New York: Palgrave Macmillan.

Foucault, Michel. 2008. *Birth of Biopolitics: Lectures at the Collège de France, 1978–79.* Translated by Graham Burchell. Edited by Michel Senellart, François Ewald and Alessandro Fontana. Basingstoke: Palgrave Macmillan.

Friedman, Milton, and Rose Friedman. 1980. *Free to Choose: A Personal Statement.* New York: Harcourt Brace Jovanovich.

Gard, Michael. 2007. "Is the war on obesity also a war on children?" *Childrenz Issues: Journal of the Children's Issues Centre* 11 (2):20–24.

Gard, Michael, and Jan Wright. 2005. *The Obesity Epidemic: Science, Morality and Ideology*. New York: Routledge.

Garland, David. 2014. "The welfare state: a fundamental dimension of modern government." *European Journal of Sociology* 55 (03):327–364.

GFKBM. 2007. *Australian Better Health Initiative Diet, Exercise and Weight.* Developmental Communications Research Report. Sydney: GFK Bluemoon.

Gordon, Colin. 1991. "Governmental rationality: an introduction." In *The Foucault Effect: Studies in Governmentality*, edited by Graham Burchell, Colin Gordon and Peter Miller. Chicago, IL: University of Chicago Press.

Grier, Sonya, and Carol A. Bryant. 2005. "Socialeditorealth." *Annual Review of Public Health* 26 (1):319–339. doi: doi:10.1146/annurev.publhealth.26.021304.144610.

Harvey, David. 2009. *A Brief History of Neoliberalism*. New York: Oxford University Press.

Hayek, Friedrich August. 2001. *The Road to Serfdom*. New York: Routledge.

Hickman, Martin. 2010. "Nudge or fudge? Public health fears as Lansley retreats from regulation." *The Independent*, December 4.

Holland, Kate E., R. Warwick Blood, Samantha I. Thomas, Sophie Lewis, Paul A. Komesaroff, and David J. Castle. 2011. "'Our girth is plain to see': An analysis of newspaper coverage of Australia's future 'fat bomb'." *Health, Risk & Society* 13 (1):31–46. doi: 10.1080/13698575.2010.540648.

Howell, Jeremy, and Alan Ingham. 2001. "From social problem to personal issue: the language of lifestyle." *Cultural Studies* 15 (2):326–351.

Ilcan, Suzan. 2009. "Privatizing responsibility: public sector reform under neoliberal government." *Canadian Review of Sociology/Revue canadienne de sociologie* 46 (3):207–234. doi: 10.1111/j.1755-618X.2009.01212.x.

Jacobson, Michael F., and Kelly D. Brownell. 2000. "Small taxes on soft drinks and snack foods to promote health." *American Journal of Public Health* 90 (6):854.

Kotler, Philip, and Gerald Zaltman. 1971. "Social marketing: an approach to planned social change." *Journal of Marketing* 35 (3):3–12.

Laczniak, Gene R., Robert F. Lusch, and Patrick E. Murphy. 1979. "Social marketing: its ethical dimensions." *Journal of Marketing* 43 (2):29–36.

Larsen, Lars Thorup. 2011. "The birth of lifestyle politics: the biopolitical management of lifestyle disease in the United States and Denmark." In *Governmentality: Current Issues and Future Challenges*, edited by Ulrich Bröckling, Susanne Krasmann and Thomas Lemke. New York: Routledge.

Lazzarato, Maurizio. 2009. "Neoliberalism in action." *Theory, Culture & Society* 26 (6):109–133. doi: 10.1177/0263276409350283.

Lefebvre, Craig R., and Flora A. June. 1988. "Social marketing and public health intervention." *Health Education & Behavior* 15 (3):299–315. doi: 10.1177/109019818801500305.

Lemke, Thomas. 2001. "'The birth of bio-politics': Michel Foucault's lecture at the Collège de France on neo-liberal governmentality." *Economy and Society* 30 (2):190–207.

Lupton, Deborah. 2004. "'A grim health future': food risks in Sydney press." *Health, Risk and Society* 6 (2):187–200.

Lupton, Deborah. 2014. "'How do you measure up?' assumptions about 'obesity' and health-related behaviors and beliefs in two Australian 'obesity' prevention campaigns." *Fat Studies* 3 (1):32–44. doi: 10.1080/21604851.2013.784050.

Marlow, Michael L. 2013. "Soda regulation is not the solution." *American Journal of Preventive Medicine* 18 (1): 14–16.

Marmot, Michael. 2005. "Social determinants of health inequalities." *The Lancet* 365 (9464):1099–1104.

Martin, Randy. 2002. *The Financialization of Daily Life*. Philadelphia, PA: Temple University Press.

McNay, Lois. 2009. "Self as enterprise: dilemmas of control and resistance in Foucault's *The Birth of Biopolitics*." *Theory, Culture and Society* 26 (6):55–77.

McSmith, Andy. 2010. "First Obama, now Cameron embraces 'nudge theory'." *The Independent*, August 12.

Ménard, Jean-Frédérick. 2010. "A 'nudge' for public health ethics: libertarian paternalism as a framework for ethical analysis of public health interventions?" *Public Health Ethics* 3 (3):229–238. doi: 10.1093/phe/phq024.

Minister for Health and Ageing. 2009. *Speech at the Australian Food and Grocery Council Dinner*. Commonwealth of Australia 2009 [cited December 23 2011]. Available from http://www.health.gov.au/internet/ministers/publishing.nsf/Content/sp-yr09-nr-nrsp281009.htm.

Mirowski, Philip. 2009. "Postface: defining neoliberalism." In *The Road from Mont Pèlerin: The Making of the Neoliberal Thought Collective*, edited by Philip Mirowski and Dieter Plehwe. Cambridge, MA: Harvard University Press.

Mont Pelerin Society. 1947. *Statement of Aims*. [cited 20/06/2010 2010]. Available from https://www.montpelerin.org/montpelerin/mpsGoals.html.

Mudge, Stephanie Lee. 2008. "What is neo-liberalism?" *Socio-Economic Review* 6 (4):703–731.

O'Hara, Lily, and Jane Gregg. 2006. "The war on obesity: a social determinant of health." *Health Promotion Journal of Australia* 17 (3):260–263.

Office of the First Lady. 2010. *Remarks of First Lady Michelle Obama – Let's Move Launch*. The White House 2010 [cited January 18 2012]. Available from http://www.whitehouse.gov/the-press-office/remarks-first-lady-michelle-obama.

Pollock, Allyson. 2005. *NHS plc: The Privatisation of Our Health Service*. London: Verso.

Public Health England. 2015. *Economics of Obesity*. Public Health England [cited 6 May 2015]. Available from https://www.noo.org.uk/NOO_about_obesity/economics.

Pusey, Michael. 2010. "25 years of neo-liberalism in Australia." In *Goodbye to All That?: On the Failure of Neoliberalism and the Urgency of Change*, edited by David McKnight and Robert Manne. Melbourne: Black.

Rayner, Geof, and Tim Lang. 2011. "Is nudge an effective public health strategy to tackle obesity? No." *BMJ* 342. doi: 10.1136/bmj.d2177.

Rose, Nikolas. 1999. *Powers of Freedom: Reframing Political Thought*. Cambridge: Cambridge University Press.

Rose, Nikolas, Pat O'Malley, and Mariana Valverde. 2006. "Governmentality." *Annual Review of Law and Social Science* 2:83–104.

Rowe, Gene, and Lynn J. Frewer. 2005. "A typology of public engagement mechanisms." *Science, Technology & Human Values* 30 (2):251–290. doi: 10.1177/0162243904271724.

Roxon, Nicola. 2008. *Australia Measures Up – National Obesity Campaign*. edited by Department of Health and Ageing. Canberra: Australian Government.

Roxon, Nicola. 2011. *Swap it Don't Stop it*. edited by Department of Health and Ageing. Canberra: Australian Government.

Secretary of State for Health. 2010a. *Healthy Lives, Healthy People: Our Strategy for Public Health in England*. [cited March 8 2012]. Available from http://www.dh.gov.uk/prod_consum_dh/groups/dh_digitalassets/documents/digitalasset/dh_127424.pdf.

Secretary of State for Health. 2010b. *Secretary of State for Health's Speech to the UK Faculty of Public Health Conference – 'A New Approach to Public Health'*. Department of Health 2010 [cited January 25 2012]. Available from http://www.dh.gov.uk/en/MediaCentre/Speeches/DH_117280.

Smith, William A. 2000. "Social marketing: an evolving definition." *American Journal of Health Behavior* 24 (1):11–17.

Snowdon, Christopher. 2013. *The Proof of the Pudding: Denmark's Fat Tax Fiasco*. IEA Current Controversies Paper 42. London: Institute of Economic Affairs.

*Sydney Morning Herald*. 2010. "Obesity – a ticking time bomb that needs to be defused." Editorial, 18 July.

Terris, Milton. 1999. "The neoliberal triad of anti-health reforms: government budget cutting, deregulation, and privatization." *Journal of Public Health Policy* 20 (2):149–167.

Thaler, Richard H., and Cass R. Sunstein. 2009. *Nudge: Improving Decisions About Health, Wealth and Happiness*. London: Penguin.

Thaler, Richard H., and Cass R. Sunstein. 2010. "Let's move". https://nudges.wordpress.com/2010/02/11/lets-move/

Thow, Anne Marie, Christine Quested, Lisa Juventin, Russ Kun, A. Nisha Khan, and Boyd Swinburn. 2011. "Taxing soft drinks in the Pacific: implementation lessons for improving health." *Health Promotion International* 26 (1):55–64.

Trinko, Katrina. 2013. "Soda ban? What about personal choice?" *USA Today*, March 10.

Truong, V. Dao. 2014. "Social marketing: a systematic review of research 1998–2012." *Social Marketing Quarterly* 20 (1):15–34. doi: 10.1177/1524500413517666.

Tsai, A.G., D.F. Williamson, and H.A. Glick. 2011. "Direct medical cost of overweight and obesity in the USA: a quantitative systematic review." *Obesity Reviews* 12 (1):50–61. doi: 10.1111/j.1467-789X.2009.00708.x.

Von Bulow, H. 2010. *How Fascism Took Over the Democratic Party*. Bloomington, IN: Authorhouse.

White House Task Force on Childhood Obesity. 2010. *Solving the Problem of Childhood Obesity within a Generation* edited by Executive Office of the President of the United States. Washington DC: The White House.

Whyte, Jessica. 2014. "Is revolution desirable?: Michel Foucault on revolution, neoliberalism and rights." In *Re-reading Foucault: On Law, Power and Rights*, edited by Ben Golder. New York: Routledge.

Wiebe, Gerhart D. 1951. "Merchandising commodities and citizenship on television." *Public Opinion Quarterly* 15 (4):679–691.

Wilkinson, Iain. 2010. *Risk, Vulnerability and Everyday Life*. London: Routledge.

Wilson, David, and William Dixon. 2012. *A History of Homo Economicus: The Nature of the Moral in Economic Theory*. New York: Routledge.

Žižek, Slavoj. 2001. *On Belief*. London: Routledge.

Zoller, Heather M. 2009. "Technologies of neoliberal governmentality: The discursive influence of global economic policies on public health." In *Emerging Perspectives in Health Communicaiton*, edited by Heather M. Zoller and Mohan J. Dutta. London: Routledge.

# 3 Lifestyle as health

## Articulating an 'urgent need'

> In the civilian world, unfit or overweight employees can impact the bottom line.
> But in our line of work, lives are on the line and our national security is at stake.
> Air Force General Richard E. Hawley, *Still Too Fat to Fight*
> (Christeson et al. 2012: 2)

Tsunamis, floods, and undermanned armies have become a cache of crisis clichés for politicians, journalists and public health advocates trying to awaken a docile public to the looming threat of obesity. These appeals are more than hyperbole or 'click-bait'. They reflect and generate a concern that obesity threatens the economy, national security, and life expectancy of the population. In the US, a group of 500 retired admirals, generals, and military leaders have formed a group called *Mission: Readiness* that believes national security is at stake due to the growing numbers of children who are purportedly too fat to serve in the US military. In 2010 they released a report *Too Fat to Fight* and a follow-up report in 2012, *Still Too Fat to Fight* (Christeson et al. 2012). Although these reports do address the influence of the school food environment on adolescent health, a dominant feature of the crisis rhetoric surrounding obesity is human agency and choice. Individual humans are responsible and therefore it is within human control to prevent and alter the predicted future of disaster.

The characterization of obesity as a threat emanating from individual choices and lifestyles establishes an *urgent need* that stimulates and enlivens the network of lifestyle. This characterization is not purely rhetorical or a pernicious neoliberal ploy, but is partly made possible via the knowledge produced by epidemiologists, nutrition scientists and public health practitioners. As discussed in Chapter 1, an urgent need is the final feature of Foucault's *dispositif* (1980: 195). Strategic networks of power, knowledge and subjectification are mobilized in response to an urgent need. It is crucial to appreciate the knowledges and relations of power used to establish a link between individual choices and obesity-as-population-threat.

This chapter sketches the intermingling of neoliberal ideas of governance with epidemiology and health policy that make the daily practices and choices of the individual visible as biopolitical targets. In locating the cause of epidemics in daily practices and choices of the individual, lifestyle epidemiology makes activities, bodies and behaviours visible for biopolitical mechanisms to target and govern. The objective of this chapter is to trace the thread of epidemiology and health promotion through the fabric of the lifestyle network. The knowledges and strategies established through epidemiology and health promotion serve to isolate obesity as an urgent need emanating from choice. First, I map the transition of explanatory models of disease from germ theory to lifestyle. Second, I trace the intersection between epidemiology and health policy, demonstrating the mutual influence of neoliberal theory and epidemiology in the creation of policies that emphasize individual responsibility for health. Finally, I discuss the subtle but significant shift in public health strategies from health education to health promotion. This shift corresponds with the transition from educating the individual to *prevent* disease, to incentivizing the individual in order to *promote* health. This chapter isolates and highlights the lines of power/knowledge in the lifestyle network that produce the conditions under which individuals are obliged to make responsible, free and healthy choices.

## From germs to lifestyle

Since the second half of the nineteenth century germ theory was the primary disease framework for health professionals, policy makers and the public. Initially health professionals and the public regarded the idea that a germ or pathogen infiltrates the human organism via the air, blood, water or food with suspicion. The story of Ignaz Semmelweis illustrates this suspicion. In the 1840s, Semmelweis recommended that Austrian physicians wash their hands using an antiseptic solution between patients. Based on an observational study that compared the rates of death from puerperal fever between two clinics, Semmelweis had concluded that higher incidence of death in one was due to the medical students and physicians moving from post-mortem to obstetric examinations without washing their hands and thereby transmitting a bacterial infection from the cadavers to the patients.[1] The suggestion that it was educated male physicians who were responsible for infecting patients, and not the midwives, was considered offensive and unfathomable. The hospital hierarchy ignored Semmelweis's recommendation. Tragically mothers continued to die from puerperal fever. The case of Semmelweis highlights the relations of power involved in the production and acceptance of epidemiological knowledge.

Despite initial scepticism (and the ruin of Semmelweis's reputation), the experiments of Louis Pasteur in the 1860s established the empirical basis of the germ theory and it became widely influential in medical and lay understandings of disease. The early findings of immunology, pathogenic theory and bacteriology of Edward Jenner, Louis Pasteur and Robert Koch resulted in the identification of disease-causing microbes and the discovery of corresponding vaccinations was made possible. Although rudimentary practices of inoculation were employed prior to the immunology and bacteriology of the late nineteenth century, the development of scientific understanding in these fields furnished epidemiology with a theoretical underpinning. The ability to identify the single causal microbe specific to certain diseases opened the possibility of preventing the spread of disease in a particular population. Germ theory proved powerful in analysing and explaining the communication of specific diseases ravaging populations. It provided the conditions to discover the bacterial or microbial cause of disease, enabling the development and rational administration of vaccinations to individuals at risk within the population.

The discovery of germ theory is a key narrative in positivist medical history. Medical historian Harry Wain argues that the initial immunology of Edward Jenner was 'more than the discovery of vaccination against smallpox, it was a basic medical discovery that changed the course of the world by introducing it to the concept of disease prevention' (1970: 187). Through the development of immunology, the communication of particular diseases and the possibility of their prevention became better understood through scientifically based epidemiology. The success in preventing communicable diseases such as smallpox, tuberculosis, polio and typhoid fuelled optimism regarding the possibility of eradicating all communicable disease. This optimism is expressed in Wain's *A History of Preventive Medicine*, written in 1970. Employing religious tropes, Wain boldly states that the medical achievement of the past is only the beginning of an 'almost unlimited promise for the future' (1970: 391). He continues,

> With the complete conquest of contagion virtually achieved, we now literally stand upon the mountain tops and look into the Promised Land. In our modern world of rapid scientific advancement, preventive medicine will leap from one great victory to another. As scientific research continues to unlock the secrets of nature, we will learn the cause and prevention of birth defects, cancer and the rest of our chronic, crippling and degenerative disease.
>
> (Wain 1970: 391)

Four decades have passed since Wain wrote these words. Rather than descend from the mountain top into the disease-free 'Promised Land', not only has the world been exposed to the horrors of HIV/AIDs, but cancer, birth defects and chronic diseases are still very much part of the medical and social landscape. Despite his prophetic shortcomings, Wain was right in pointing to the shift in focus from contagion to 'diseases of lifestyle'.

The post-war affluence experienced in many Western societies combined with the success of preventing a number of infectious diseases contributed to a decrease in early death rates and altered the demographic makeup. Public health historian Milton Lewis observes that by the second half of the twentieth century, the 'populations of Australia, Britain and America were ageing, and chronic, "degenerative" diseases were replacing infectious diseases as the main causes of death and disability in the affluent societies' (2003: 2). Cancers, heart disease, and diabetes replaced smallpox, tuberculosis and cholera as the maladies of Western society. This shift challenged both epidemiology and prevention strategies. The search for a micro-organism as the sole cause for cancer or heart disease proved unsuccessful. Not only was there an increased 'difficulty in identifying aetiological agents' of chronic diseases, some researchers began to acknowledge that 'many have multiple causes' (Farmer and Lawrenson 2004: 92, Galdston 1954).

The transition in epidemiology from microbes to lifestyle ushered in a new era for public health and disease prevention. Milton Terris, former President of the American Public Health Association (1966–67), suggests that the work of epidemiologists to link chronic and non-infectious disease to lifestyle introduced a second epidemiologic revolution (1992: 186). This revolution mobilized techniques of health education, taxation and regulation in service of Public Health's 'major battles' and 'struggle against unhealthy lifestyles' (Terris 1992: 188). However, the focus on lifestyle to explain disease assumes that certain diseases are the result of individual choices and simplifies complex 'interrelationships between many variables in disease aetiology' (Hansen and Easthope 2007: 7). No longer was a single microbe the cause of disease, but the many variables of an individual's daily habits and choices.

Despite acknowledgements of multi-causal disease aetiologies, the lifestyle theory of disease causation is not completely disentangled from the germ theory. Both draw on a biomedical conception of causation that tries to explain the whole (incidence of disease in a population) by reducing it to the parts (individual cases) (Krieger 2011: 136). According to Geoffrey Rose, this is an erroneous step as the 'determinants of incidence are not necessarily the same as the causes of cases' (2001a: 429). A public health strategy that targets the causes of cases (individual behaviours) rather than the causes of incidences (systemic and population-wide factors) will be ineffective or overly burden individuals. Rose and others who were aware of the lacunae between individual cases and population incidences suggested that governments needed to address structural determinants of health not simply individual behaviours (Krieger 2011). However, the political will and lifestyle theory of disease in the mid-twentieth century was not receptive to research suggesting that social inequality and class conditioned the fortunes of individuals and populations. Instead, 'individually-oriented theories of disease causation, in which population risk was thought to reflect the sum of individuals' risk' (Krieger 1994: 890) appealed to popular ideas of individualism and found support in government health departments influenced by neoliberal ideas (Mayes 2014).

Notwithstanding theoretical and practical criticisms, lifestyle epidemiology became the dominant theory of disease from the 1960s onwards. Hansen and Easthope identify four features that establish a connection between lifestyle and health: choice, behaviour, evidence and risk factors (2007: 17–19). First, lifestyle is regarded as a matter of choice and therefore considered adjustable and amendable. Second, following a biomedical path, behaviours associated with health status (diet or smoking) are isolated as causal agents, rather than those systemic factors such as socio-economic status. Third, the spectrum of behaviours that impact health are narrowed to those that can be scientifically and statistically 'shown to impact (usually negatively) physical health' (Hansen and Easthope 2007: 18). Rather than understanding health as wellbeing, resulting from a wide spectrum of social interactions, lifestyle epidemiology is prejudiced toward the 'biomedical understanding of health that is mechanistic and biological' (Hansen and Easthope 2007: 18). Fourth, through statistical studies of the distribution of disease within a population, lifestyle epidemiologists attempt to isolate risk factors associated with a particular disease, for example, an association at the population level between the consumption of saturated fat and incidence of coronary heart disease. The consumption of saturated fat is then considered a lifestyle risk factor that may increase the probability of developing heart disease (Hansen and Easthope 2007: 19). Importantly, lifestyle epidemiologists regard these risks as the 'lifestyle choice' of the individual and it becomes therefore the responsibility of the individual to modify behaviours through correct (i.e. healthy) choice.[2] These four features of lifestyle epidemiology produce an explanatory framework for both health promotion and disease prevention.

The concern over lifestyle provides the conditions for health promotion. Rather than preventing disease, health promotion is a distinct and unique approach to public health. Washing ones hands or using a condom prevents communicable disease, but they don't promote health. Eating vegetables or jogging, however, are characterized as both disease preventing and health promoting. The lifestyle approach is used to direct individuals to maximize their health through choices. Lifestyle forms a structure that organizes individual behaviours as determining health, transforming the relationship between the individual and health from passive to active. Further, empirical sociological research demonstrates that lay understanding of lifestyle is used to retrospectively explain cases of disease, assess present health status and predict future health or disease (Davison, Frankel, and Smith 1992, Davison, Smith, and Frankel 1991). The lifestyle rhetoric reconceives health from a passive state that is the result of chance, providence or biological luck, to a fluctuating goal that is within an individual's control yet requires continual work and promotion in order to maintain it. The idea that health can be promoted and fortified by individual choice is taken up and reinforced by neoliberal health policy.

## Lifestyle and the rhetoric of neoliberal health policy

The account of lifestyle epidemiology outlined here does not exhaust the theoretical and practical features of epidemiology. My purpose has not been to provide a comprehensive overview of epidemiology, but to demonstrate the way select aspects of lifestyle epidemiology are used in political discourse as a means to justify policies focusing on the individual. The dissemination of lifestyle epidemiological ideas through public health campaigns, social policies and media reports shape public understanding of the relationships connecting disease, health and individual choice. In the 1970s and 80s, certain ideas found within lifestyle epidemiology were used by politicians to introduce health policies with the dual objective of emphasizing individual freedom and cutting health expenditure (Irwin and Scali 2007: 245). The rhetoric of anti-obesity campaigns and policy announcements continue to draw heavily on the idea that health and disease are predominately determined by individual choice and behaviour. The policies introduced in the late 1970s provided the policy paradigm through which contemporary governance of obesity is formed (Hall 1993). To trace this development I briefly address aspects of the neoliberal political rhetoric of lifestyle in Australia, the US and the UK.

The political and economic conditions of the 1970s in much of the West encouraged the incorporation of lifestyle epidemiology into public health policy. Confronted with an increase in chronic disease that could not be explained or addressed through the germ theory, health policy makers in the 1970s appealed to the explanatory framework of lifestyle. The germ theory was popular with policy makers during the period of the welfare state, as it required little 'personal change or economic upheaval' (Hansen and Easthope 2007: 2). Preventive techniques such as quarantine or vaccinations were relatively cheap, effective and operated with ideas of utility or socialism that allowed centralized State bureaucracies to administer them. In contrast, neoliberal ideas of the free market and non-interference harmonized with lifestyle epidemiology to focus on the freedom, choice and responsibility of the individual. I am not suggesting that economic theory determines epidemiology, but that there is a relationship of efficiency between them as they allow for smooth interaction at the level of policy and implementation.

The epidemiological concept of lifestyle came to prominence in political and health policy parlance through Canadian Health and Welfare Minister Marc Lalonde. In 1974, Lalonde produced *A New Perspective on the Health of Canadians*, commonly referred to as the Lalonde Report. Although it focused on the Canadian health system, the Lalonde Report had an international impact (Lewis 2003: 151). The report 'describes medicine and health care services as one of the four "health fields" that influenced health and illness, the others being human biology, the environment and lifestyle' (Baum 2003: 32). The idea that disease is the result of multiple fields rather than a single germ, gene or activity presented a practical challenge to policy makers and health professionals. The multiple cause theory 'gave few clues as to "how" to prevent disease and promote health' (Goltz and Bruni 1995: 527). If the environment, biology, health services and individual lifestyle contribute to disease then there are an infinite number of causes and influences on health. Most of these causes were considered as out of the control of the State, for economic and ideological reasons. If they were within State control, the required change was usually seen as either too costly or too complicated. Owing to the influence of neoliberal ideas of decentralization and non-interference, and using epidemiology as a justification, Australia, the US and the UK introduced public health strategies focusing on individual behaviours.

Despite the Lalonde Report's emphasis on environment and socio-economic status as influencing health outcomes, lifestyle was the prominent theme accepted by policy makers. For example, on 25 May 1979 the Australian Federal Minister for Health, Ralph Hunt, stated:

> During my period of office as Minister for Health I have become more and more convinced that continued concentration on traditional curative medicine, with its associated high costs both for the Government and the

individual, can add little to improving the nation's health status. *I believe this can be achieved only by motivating individuals to take a responsible attitude for their own personal health care.*

(Hetzel 1980: 256) (Emphasis added)

The primary role for the government, according to Hunt, is to motivate individuals and provide the conditions for individuals to adopt a 'responsible attitude' for their own health.

The rhetoric of individual responsibility and State withdrawal is also found in the United States. Following from the US Surgeon General's 1979 report, *Healthy People,* Richard S. Schweiker, the Secretary of Health and Human Services (1981–1983), outlined the Reagan administration's strategy for prevention and health promotion. Schweiker believed that Americans are 'interested in exercising some responsibility and control over our own health' (1982). And given the correct information individuals would make the 'simple personal efforts' necessary to reduce their risk of lifestyle diseases (Schweiker 1982: 197–198). Fortunately for Schweiker, making the correct information available to the public coincided with the neoliberal ideas of the Reagan administration and did 'not require massive infusion of Federal funds' (1982: 197). Instead of federally-funded services or infrastructure, Schweiker believed a combination of health education programmes and private sector initiatives would provide the necessary opportunities 'to change lifestyle, to change habits, and to make people live healthier and longer' (1982: 197). Echoing Reagan's state-phobic rhetoric, Schweiker emphasized that 'Washington alone cannot persuade Americans to incorporate prevention into their lifestyles', but through education and effort 'most people can make daily decisions that influence their health' (1982: 197–198). Under Schweiker and Reagan, lifestyle theory was established as 'the centrepiece of U.S. chronic disease prevention policy' (Tesh 1988: 45).[3]

Running parallel to the Australian and US contexts, in the UK the Thatcher Government introduced lifestyle focused health policies. However, prior to the victory of Thatcher's Conservative Party in 1979, the Labour government commissioned Douglas Black to write the *Report of the Working Group on Inequalities in Health* (Black et al. 1980). Completed in 1980, the report recommended 'equity-oriented social policy measures in areas such as education, housing, and working conditions, as well as modifications in the delivery of health care services' (Irwin and Scali 2007: 245). The Report stated that these measures would assist in narrowing existing gaps in health status between different socio-economic groups. When presented to the Thatcher Government the recommendations to address social determinants of health were rejected for being too reliant on government intervention into the lives of individuals, society and the market (Irwin and Scali 2007). Instead the Secretary of State for Health and Social Security, Patrick Jenkin (1979–1981), stressed the importance of individual responsibility. According to Jenkin '[t]he doctrine that somehow it is the responsibility of the authorities to care for us from the cradle to the grave so that we have no responsibility for our necessities or for tribulations in life is so totally contrary to human nature' (1981: 240–241). The idea that the State should be responsible for the welfare of the population is, according to Jenkin, the 'complete antithesis of what most of us mean by a free society' (1981: 241), and therefore individuals, their families and communities were increasingly expected to assume responsibility and control of their lives and health.

This policy rhetoric was not simply accepted without debate or protest, however the idea that individuals are responsible for the health consequences of their choices became a bipartisan feature of health policies from the 1970s to the present. Australian, US and UK Health Ministers and governments shaped policy parameters in order to remove the burden of disease prevention and health maintenance from the State and place the responsibility on the individual (Terris 1999, Irwin and Scali 2007, Bury 1994). The individual became the *only* actor able to reduce the economic and social costs of chronic disease. This view was not only in accordance with epidemiological knowledge, but also with what it purportedly meant to be a free human being.

In addition, the individual was positioned as not only responsible for securing their own health, but also securing the *nation's health status*. The task that remained for the State was to provide information and incentives to ensure that individual behaviour change was achieved. Lifestyle focused education campaigns sought to inform and enable individuals to make responsible, healthy and rational choices (Cockerham 2005: 53). Educating individuals about diet, exercise and daily habits to prevent the development of chronic disease became a central public health tool.

## Health education and the new public health

Health education has been an important tool of public health since at least the medical and social hygiene movement of the nineteenth century.[4] In this context, education revolved around behaviours associated with transmission of germs. However, as the causal agent of disease increasingly moved from microbes into the sphere of the individual's everyday life, the importance and use of health education strategies intensified (Hansen and Easthope 2007: 15). As demonstrated in the previous chapter's analysis of *Measure Up*, educating the public about health statistics and risk indicators is a key component in strategies to alter behaviours associated with obesity.

These education strategies imply and depend on a conception of the individual as a rational and free chooser. That is, given the correct information the autonomous and self-determining individual will make the free and responsible choice that promotes their own health and the health of the population. Put simply, and it often is, if individuals are educated to know that smoking causes cancer or sugar causes diabetes then they will change their behaviour.

The focus on education has not been without its critics. Libertarians and commentators concerned with 'healthism' believe the shift towards lifestyle represented illegitimate government overreach into the everyday lives of individuals (Crawford 1980, Skrabanek 1994). They argue that the focus on lifestyle does not mark the transformation of the welfare state into a neoliberal state, but serves as an avenue for governments to establish 'greater surveillance over their population's lifestyles and behaviour' (Bury 1994: 24). I addressed the healthism critique and its relation to my argument in Chapter 1. We do not need to revisit it here, except to re-emphasize two inadequacies. First, lifestyle is not a strategy created by a sovereign (the state) that is intentionally applied to a population for a specific political end. The lifestyle network is 'a strategy without a strategist'[5] comprised of an entanglement of threads (epidemiology and health policy) that operate in an unpredictable manner. Second, as will be made clearer in following chapters, lifestyle is a network in which the individual is actively engaged and in which they use their freedom to choose, adopt and fashion a healthy lifestyle.

In addition to the libertarian or healthism critique there is a second line of criticism directed at the lifestyle approach to public health – the 'new public health'. The *new public health* draws on the Lalonde Report, yet attempts to incorporate the three areas marginalized by governments: environment, health services and biology. Practitioners, researchers and advocates associated with the new public health perspective argue that individual-focused education strategies fail to account for the impact of social, environmental and infrastructure factors on the ability of the individual to freely choose and control lifestyle. The new public health emphasizes health promotion as an empowering approach to public health that does not solely focus on individual behaviour change.[6]

The discourse of health promotion represents an important development in the use of lifestyle as a strategy to secure population health. Following the Lalonde Report, a second document that had an international impact on the use of lifestyle in public health was the 1986 World Health Organization's 'Ottawa Charter for Health Promotion'. In an apparent critique of neoliberal approaches focusing on individual lifestyles, the Ottawa Charter proposes five strategies to achieve health promotion: the development of healthy public policy; the creation of supportive environments; strengthening community action; the development of personal skills; and reorientation of health services (Baum 2003: 34). The Ottawa Charter's definition of health and health promotion is worth quoting in full. The charter states:

> Health Promotion is the process of enabling people to increase control over, and to improve, their health. To reach a state of complete physical, mental and social well-being, an individual or group must be able to identify and to realize aspirations, to satisfy needs and to change or cope with the environment. Health is therefore seen as a resource for everyday life, not the objective of living. Health is a positive concept emphasizing social and personal resources, as well as physical capabilities. Therefore, health promotion is not just the responsibility of the health sector, but goes beyond healthy life-styles to well-being.
>
> (World Health Organization 1986)

Health rather than disease is the clear focus. Health is used seven times, while disease is not mentioned. The shift from preventing disease to promoting health is reflected in entrepreneurial and market-oriented notions like 'resource', 'improvement', 'control' and 'enablement'. This language suggests that the Ottawa Charter is not as far removed from the neoliberal policy milieu as some proponents may believe. While many public health advocates were disappointed with the individual focus of the lifestyle approach stemming from the Lalonde Report, the Ottawa Charter 'captured the imagination of the health promotion community in Western countries' (Lewis 2003: 151). According to Baum, the Charter became 'the new public health Bible' as it established the idea that population health it is not secured through healthy choices of individual (2003: 34), but through 'government policies that change the structures people live, work and play in' (Baum and Sanders 2011: ii253). The Ottawa Charter stresses the influence of factors beyond an individual's control or choice. However, to the frustration of some commentators, the Charter's emphasis on personal skills was seen as maintaining the lifestyle and behavioural approaches to health promotion (Baum 2003: 34–35, Goltz and Bruni 1995: 510). While the inclusion of personal skills as a determiner of health was received with suspicion by some within the public health community, others regarded the integration of these tensions within the Ottawa Charter as a sign of the document's strength (Beaglehole and Bonita 2004: 255–256).

The new public health sought to continue and re-establish the political aspect of public health, exemplified by eighteenth- and nineteenth-century reformers such as Jeremy Bentham, Edwin Chadwick and Friedrich Engels,[7] concerning 'itself with social causes of disease, health status inequalities, human rights and global environmental issues' (Lewis 2003: 10). However, unlike the wider critique of political economy offered by Bentham, Chadwick and Engels, the conceptual language of the Charter and health promotion is arguably attuned to rather than critical of neoliberal political rationality.

The Ottawa Charter's emphasis on the individual, particularly through notions of empowerment and choice, exposed health promotion and the new public health to criticisms. According to Russell Caplan, both the 'educational and self-empowerment approach have the *individual* as the effective unit on which to target health education/promotion initiatives' (1993: 147). Whether disease is prevented through education or health is promoted through empowerment, the individual remains at the centre. The conception of health as a resource to be promoted merges neatly with the neoliberal conception of the freely choosing individual as well as the commodification of health, social security and everyday life. Karen Goltz and Nina Bruni argue that health promotion and the new public health represent a language change rather than a fundamental shift from the health education and lifestyle focus of the 1970s. For example, they suggest that the 'new discourse of health promotion constitutes "enablement" rather than modification or communication as the desired approach to be employed in the recruitment of subjects to new public health cause' (Goltz and Bruni 1995: 530). Rather than educating individuals to modify behaviours, the health promotion strategy employs incentive-based regulations to enable the individual to invest in, control and use their health as a resource. Victoria Grace is cited in Goltz and Bruni as arguing that the health promotion model of health depends on a 'world of production and consumption operating in accordance with incentive and disincentives, with measurable inputs and outcomes, and a discourse of rational individuals using cost-benefit decision-making' (1995: 530). Rather than 'liberating' the individual from the 'oppression' of lifestyle-focused health education, the shifts from disease to health, education to promotion, and modification to enablement represents an increasingly tighter entanglement of the individual in the network of lifestyle, further enabling biopolitical governance.

As demonstrated by the rhetoric of the politicians who in the 1970s mobilized lifestyle as the cause and cure of health problems, individuals are expected to control and determine their own health outcomes. The biopolitical rationality operating in lifestyle and new public health strategies stresses individual responsibility for establishing a healthy lifestyle that conforms to biomedical norms despite structural forces that may or may not influence these choices. Rather than abandoning or broadening the focus on the individual, health promotion strategies 'are underpinned by ideologies of individualism, autonomy, self-determination and rationality' (Hansen and Easthope 2007: 145–146). The individual is situated in a broader biopolitical milieu requiring the negotiation and enablement of lifestyle choices. Through this approach, health increasingly is expanded and enmeshed in all aspects of the individual's social, psychological and biological life (Engel 1977). Thus health is transformed from a negative state of 'not being ill' into a positive resource that one should actively seek to develop for oneself and for the security of the population (Hansen and Easthope 2007: 145–146). The lifestyle network makes a healthy lifestyle a visible indicator of the neoliberal subject's success or failure to take responsibility for their own future and that of their family and society.

## Hiding the violence of norms in lifestyle

In addition to shifting the burden of responsibility to individuals, health promotion circulates the norms of health that identify and make visible noncompliant individuals. Writing in the early 1980s, epidemiologist Geoffrey Rose addresses the difficulty for individuals who 'step out of line' with norms of unhealthy eating, drinking and smoking. According to Rose, '[i]f we try to eat differently from our friends it will not only be inconvenient, but we risk being regarded as cranks or hypochondriacs' (2001a: 431). However, the health norms of the second decade of the twenty-first century represented in media and public health have arguably shifted toward those of the 'crank' and 'hypochondriac'. Smoking cessation and eating in accordance with norms of nutritional health have become desirable (Howland, Hunger, and Mann 2012). This is not to suggest that individuals always act in accordance with these norms, but that they recognize and circulate the value of the norms (Croll, Neumark-Sztainer, and Story 2001). In this context, individuals that make choices that 'step out of line' of norms become visible as objects requiring direct instruction, disciplinary measures or exclusion from the population.

In Chapter 1, I introduced the biopolitical impulse of care and violence that makes live and lets die. This impulse operates in strategies of health promotion that implicitly group individuals into categories of healthy and unhealthy, compliant and deviant, or in the terms of the pastor – sheep and goats. The violence towards noncompliant individuals is not subjective (i.e. a physical beating) but exclusionary and systemic. Systemic violence exposes the individual to more disciplinary interventions that *foster* or *disallow* life. This violence is often associated with economic, social status and race.

In the case of obesity, individuals are encouraged to exercise regularly and eat healthful food to achieve the norms of a 'healthy body'. Failure to achieve and maintain a 'healthy body', usually determined by BMI or waist measurements, indicates a failure to follow lifestyle guidance and thereby legitimates increased medical and non-medical interventions. Examples include: surveillance of the home and parenting techniques (Murtagh and Ludwig 2011, Rudolf 2011, Lowe 2012), body-modifying and potentially life-threatening surgery (Omalu Bi and et al. 2007, O'Brien, Brown, and Dixon 2005), or cuts to welfare and sick pension (Boseley 2015). In addition to medical interventions, the failure to comply with body norms can expose individuals to a range of discriminatory practices: reduced employment prospects, decreased remuneration or higher insurance premiums (Puhl and Brownell 2001, Leonhardt 2009, Bhattacharya and Bundorf 2005). Lily O'Hara and Jane Gregg write,

> Evidence of systematic bias against people of higher-than-average body weights has been found in health workers, health promotion practitioners, doctors, nutritionists, coaches, employers, landlords and teachers, and in all settings including hospitals and general practices, workplaces, schools and universities.
>
> (O'Hara and Gregg 2006: 261)

These biases influence the health outcomes of individuals outside the norm and reinforce exclusion from the normative group. The conception of the individual as a free and rational chooser that only requires correct information to choose the healthy choice does not account for the influence of systemic violence on the capacity to *freely* choose.

Systemic violence, the violence inherent in the normal order of things (Žižek 2009: 2), is manifest in the continual rejection of research demonstrating the social determinants of health. For instance, Jenkin and Thatcher rejected the Black Report in 1979 and Andrew Lansley and David Cameron marginalized the Marmot Review in 2010 (Marmot 2010). Both reports identified social and structural factors as the main determiners of health. These reports undermine the belief that individuals can control their own health by simply following health education advice (Lewis 2003: 45, Davison, Frankel, and Smith 1992: 68). An individual may desire to adopt what is called a healthy lifestyle but the cost or availability of fresh food may prohibit healthful eating, and the lack of leisure time and safe and affordable recreation facilities may exclude them from engaging in physical activity. Therefore systemic factors limit an individual's ability to conform to health norms (Cockerham 2005).

There is not only a violence in the structures that limit choice, but also in the characterization of what is considered a 'correct choice' and a 'healthy body'. While the social determinants of health literature reveals important features of systemic violence inherent in the setup of society, there is a further aspect requiring discussion – the norms of health. Race, gender and economic discrimination in the allocation of health care services are significant problems that influence an individual's capacity to *choose* health (Sorlie et al. 1992, Lowe, Kerridge, and Mitchell 1995, Gittelsohn, Halpern, and Sanchez 1991). However, race, gender and economics also influence the very norms that individuals are expected to choose and adopt. For example, the norms of health that are often considered stable and scientifically objective have been historically constituted using white, male and middle-class populations. According to Hansen and Easthope 'white adult men rather than adult women or people from different racial or ethnic backgrounds' provide the population data sets from which norms of health and disease are established (2007: 36). Incidences of illness among ethnic and working class populations are often attributed to their customs, practices and styles of life, rather than intergenerational discrimination and social policies that limit education and employment opportunities as well as access to healthcare services. In contrast, health among the white and middle-class populations was credited to superior biology and way of life, rather than social policies and historical privileges that offer a wider range of opportunities.

For example, 'soul food' is an African-American cuisine deeply entwined with cultural history of slavery and the civil rights movement in the 1960s (Shute 2012). However, some health experts blame soul food for contributing to the rise in incidences of obesity, diabetes and heart disease among African-American communities (James 2004). Soul food is characterized by dishes high in fat, salt and sugar. Soul food is not beyond modification. Like all cuisines, soul food has been, and will continue to be, modified in response to certain contingencies, including nutritional guidance (Oldways Preservation Trust 2013, Barclay 2011). However, focusing on soul food to explain incidence of disease among African-American communities in America can also mask the effects of social and economic influences on health and wellbeing.

In describing soul food as 'unhealthy' and the culprit for disease among African-American communities, it is important to question whether there are values beyond nutritional health at play. Tony Whitehead notes a 'lack of comprehensive, multidisciplinary approaches that consider the cultural meaning of food as well as its biomedical consequences' not only undermines cultural values but also leads to ineffective public health interventions (1992: 95). Soul food becomes the easy target to address, rather than the continuing effects of racism, inadequate public infrastructure, educational disadvantage and limited employment opportunities among these communities. That is, black culture and black lifestyles are the problem, not systemic racism and socio-economic structures.

There has been a long history of associating health status with certain conceptions of race or ethnicity. In the late nineteenth century the white middle-class lifestyle was established as the norm whereas the '[w]orking-class or ethnic cultural lifestyles were by definition seen as unhealthy' (Hansen and Easthope 2007: 9). Perhaps no longer as overt, these norms continue to operate through the emphasis on behaviour modification resulting from health education. The lifestyle practices of the middle class professional become the practices to which the rest of the population is encouraged to replicate – shopping at farmers' markets, jogging, yoga and cycling (Petersen and Lupton 1996, Gillick 1984, Guthman 2011a). In addition, the conception of the individual capable of responding to the education campaign serves as the norm to which all other individuals and populations are required to conform. The self-directing and self-governing individual exposes those unable to adhere to health education as irresponsible and lacking discipline (Glassner 1989, Lupton 1995).

There is considerable debate over biological concepts of race (Cooper and David 1986, Rose 2001b: 155, Gilman 2008: 102). I do not intend to canvass that debate, but highlight the gendered and racial biases implicit in the norms of health and the body (Guthman 2011b: 28, Gard and Wright 2005: 60, Bordo 2003: 34). The systemic violence inherent in health norms and the health education strategies targeting lifestyle serves not only to produce healthy subjects, but also to reproduce norms associated with economic, social and racial status. In the case of *Measure Up*, population based lifestyle recommendations are grounded in health research developed for 'Caucasian' adults. The use of norms specific to the 'Caucasian ethnicity' as the basis for the population norm is acknowledged in the *Measure Up* campaign materials. A caveat in small print states:

> The waist measurements above are recommended for Caucasian men, and Caucasian and Asian women. Recommended waist measurements are yet to be determined for all ethnic groups. It is believed that they may be lower for Asian men than for Caucasian men and are likely to be higher for Pacific Islanders and African Americans (men and women). The limited data currently available indicates that the risk factors in Aboriginal populations appear to be similar to those in Asian populations; and the risk factors in Torres Strait Islander populations appear to be similar to those found in Pacific Islander populations.
>
> (Australian Better Health Initiative 2007: 3)

Consistent with this qualification, the print and television commercials feature ethnically 'Caucasian' models with the waist norms of 94 cm for men and 80 cm for women. However, it is problematic that there is no mention of the research base for these measurements. Despite the white nuclear family serving as the model, the scientific measurements used in the campaign are explicitly extended to all Australians. The message of the campaign is that obesity is a problem for all Australians and those who fall outside of these measurements need to work on their bodies and modify choices to conform to the norm. With norms functioning to bring individuals in line with the population, the 'Caucasian' specific norms suggest that this is the population that is cared for and those unable to conform are a threat and, by definition, excluded from the population. In other words, if all Australians need to conform to a specific body norm, as the campaign suggests, then failure to conform to that norm places a question mark over an individual's 'Australian-ness' and jeopardizes their position within the population.

In addition, the research base for the social marketing campaign (discussed in the previous chapter) also subtly excludes people based on social class, economics and race. The focus groups conducted by GFK Bluemoon found that 'Aboriginal and Torres Strait Islander participants recognised that lifestyle related chronic disease is a particular problem in their communities' (GFKBM 2007: 7). However, the 'enormous structural barriers to change and distrust of government advice' meant that a social marketing campaign would be an inadequate response (GFKBM 2007: 7). The focus group revealed that Aboriginal and Torres Strait Islander people, new migrants, and blue collar workers were at highest risk of chronic disease and disproportionately represented among the 'Defiant Resisters', 'Quiet Fatalists' and 'Help Seekers' segments. Despite being most in need of health and welfare service, the approach on offer – a lifestyle-focused social marketing campaign – would inadequately addresses these needs. Rather than re-evaluate the objectives of the programme to align with these findings, the $500 million *Measure Up* campaign was launched to target those able to make 'healthy choices' unencumbered by structural barriers. GFK Bluemoon recommended that in the future the government should 'explore other creative means of enhancing appreciation of "why", with the aim of engaging "Defiant Resisters" and "Quiet Fatalists", including those in Aboriginal and Torres Strait Islander communities' (GFKBM 2007: 8).

Further analysis of the operation of norms associated with race and economic status in the lifestyle network is necessary, however, from the campaign material it is evident that the norms of health and the body are not something to which everyone within the Australian population is able to conform. The urgent need to secure the population from the threat of those outside biomedical norms of health serves to activate the lifestyle network as an exclusionary mechanism. Behaviours, bodies and choices are scrutinized and assessed in comparison to scientifically determined measurements, socially recognizable cues and politically influenced forms of life. This reveals not only who is 'included' and who is aspiring to be 'included', but also who is 'excluded'.

The norms used in lifestyle strategies may not necessarily be adapted to *actual* behaviour. Empirical studies suggest that individuals adopt, appropriate and reject different aspects of lifestyle advice (Davison, Smith, and Frankel 1991, Bury 1994, Backett, Davison, and Mullen 1994). A gap between the rhetoric and practice of lifestyle does not imply a weakness of the lifestyle network as a mechanism of biopolitical governance, but part of its success. The emphasis on lifestyle and individual choice in the 1970s to curb chronic disease has not been abandoned or discredited due to the failure of individuals to *actually* adopt lifestyle advice. Despite a failure of compliance and that recommendations are 'at best a distortion of the epidemiological evidence' (Davison, Smith, and Frankel 1991: 16), there has not been a diminishment of the zeal for the lifestyle approach. Rather, the lifestyle approach has been repeated and reinforced as the appropriate strategy to prevent disease and promote health.

The 'successful failure' of lifestyle approaches recalls Foucault's conclusion that the prison successfully produces delinquency rather than rehabilitates criminals (Foucault 1991: 277). The 'success' of the lifestyle approach is not in getting individuals to adopt healthy lifestyles, but articulating an urgent need and exposing individuals who do not conform to the behavioural and health norms that secure population health.[8] Individuals actively seek to shape their daily lives to represent responsible conformity to norms. A significant part of this process is the line of subjectification in the enabling network. The following chapter addresses the way lifestyle in consumer societies is used to form identity and group belonging.

## Notes

1  The second clinic served as a control group as it was not used for educating doctors and did not have a morgue attached (Wyklicky and Skopec 1983).
2  For critical analyses of the science base of these studies see (Mayes and Thompson 2014, Krieger 2011, Rothstein 2003).
3  For a detailed analysis of the actual US policies and regulations, and how they do not necessarily map on to the public political rhetoric, see (Oberlander 2003).
4  Jacques Donzelot writes that the medical hygienists would 'employ the state as a direct instrument, as a *material means* for averting the risks of a destruction of society through the physical and moral weakening of the population' (Donzelot 1980: 56).
5  On this idea see (Dreyfus and Rabinow 1983: 187, Foucault 1998: 94–96).
6  For a critical analysis of the new public health see (Petersen and Lupton 1996).
7  For a discussion of the way the 'new public health' positions itself in relation to the philosophical and social reformers stemming from the Enlightenment see (Rosen 1993: 110–111) (Petersen and Lupton 1996) and (Baum 2003).
8  I would like to acknowledge Lars Thorup Larsen for drawing my attention to the similarities between the failure of the modern prison and the failure of lifestyle approaches.

## References

Australian Better Health Initiative. 2007. *Time to Take Some Healthy Measures? How Do You Measure Up?* Australian Government [cited August 15 2009]. Available from http://www.australia.gov.au/MeasureUp.

Backett, Kathryne, Charlie Davison, and Kenneth Mullen. 1994. "Lay evaluation of health and healthy lifestyles: evidence from three studies." *British Journal of General Practice: The Journal of the Royal College of General Practitioners* 44 (383):277–280.

Barclay, Eliza. 2011. *How Soul Food Can Be Good For Your Health.* [cited July 31 2013]. Available from http://www.npr.org/blogs/thesalt/2011/11/10/142207019/how-african-americans-can-get-healthy-with-big-helpings-of-soul-food.

Baum, Frances Elaine. 2003. *The New Public Health.* Singapore: Oxford University Press.

Baum, Frances Elaine, and David M. Sanders. 2011. "Ottawa 25 years on: a more radical agenda for health equity is still required." *Health Promotion International* 26 (suppl 2):ii253–ii257. doi: 10.1093/heapro/dar078.

Beaglehole, Robert, and Ruth Bonita. 2004. *Public Health at the Crossroads: Achievements and Prospects*. Second edn. Cambridge: Cambridge University Press.

Bhattacharya, Jay, and M. Kate Bundorf. 2005. *Incidence of the Healthcare Costs of Obesity*. National Bureau of Economic Research Working Paper Series 11303. Cambridge, MA: NBER.

Black, Douglas, J.N. Morris, C. Smith, and P. Townsend. 1980. *Report of the Working Group on Inequalities in Health*. London: Stationery Office.

Bordo, Susan. 2003. *Unbearable Weight: Feminism, Western Culture, and the Body*. Berkeley, CA: University of California Press.

Boseley, Sarah. 2015. "David Cameron's plans for obese benefit claimants questionable, says the Lancet." *The Guardian*, 19 March.

Bury, Michael. 1994. "Health promotion and lay epidemiology: A sociological view." *Health Care Analysis* 2 (1):23–30. doi: 10.1007/bf02251332.

Caplan, Russell. 1993. "The importance of social theory for health promotion: from description to reflexivity." *Health Promotion International* 8 (2):147–157.

Christeson, William, Amy Dawson Taggart, Soren Messner-Zidell, Mike Kiernan, Judy Cusick, and Ryan Day. 2012. *Still Too Fat to Fight*. Washington DC: Mission Readiness.

Cockerham, William C. 2005. "Health lifestyle theory and the convergence of agency and structure." *Journal of Health and Social Behavior* 46 (1):51–67. doi: 10.1177/002214650504600105.

Cooper, Richard, and Richard David. 1986. "The biological concept of race and its application to public health and epidemiology." *Journal of Health Politics, Policy and Law* 11 (1):97–116. doi: 10.1215/03616878-11-1-97.

Crawford, Robert. 1980. "Healthism and the medicalization of everyday life." *International Journal of Health Services: Planning, Administration, Evaluation* 10 (3):365–388.

Croll, Jillian K., Dianne Neumark-Sztainer, and Mary Story. 2001. "Healthy eating: what does it mean to adolescents?" *Journal of Nutrition Education* 33 (4):193–198. doi: http://dx.doi.org/10.1016/S1499-4046(06)60031-6.

Davison, Charlie, George Davey Smith, and Stephen Frankel. 1991. "Lay epidemiology and the prevention paradox: the implications of coronary candidacy for health education." *Sociology of Health & Illness* 13 (1):1–19. doi: 10.1111/j.1467-9566.1991. tb00085.x.

Davison, Charlie, Stephen Frankel, and George Davey Smith. 1992. "The limits of lifestyle: Re-assessing 'fatalism' in the popular culture of illness prevention." *Social Science & Medicine* 34 (6):675–685. doi: 10.1016/0277-9536(92)90195-v.

Donzelot, Jacques. 1980. *The Policing of Families*. London: Hutchinson.

Dreyfus, Hubert, and Paul Rabinow. 1983. *Michel Foucault: Beyond Structuralism and Hermeneutics*. 2nd edn. Chicago, IL: University of Chicago Press.

Engel, George L. 1977. "The need for a new medical model: a challenge for biomedicine." *Science* 196 (4286):129–136.

Farmer, Richard, and Ross Lawrenson. 2004. *Lecture Notes: Epidemiology and Public Health*. Fifth edn. Oxford: Blackwell Publishing.

Foucault, Michel. 1980. "The confession of the flesh." In *Power/Knowledge: Selected Interviews and Other Writings*, edited by Colin Gordon. New York: Pantheon Books.

Foucault, Michel. 1991. *Discipline and Punish: The Birth of the Prison*. Translated by Alan Sheridan. London: Penguin.

Foucault, Michel. 1998. *The Will to Knowledge: The History of Sexuality Volume 1*. Translated by Robert Hurley. Harmondsworth: Penguin Books.

Galdston, Iago, ed. 1954. *Beyond the Germ Theory: The Roles of Deprivation and Stress in Health and Disease*. New York: Health Education Council, New York Academy of Medicine.

Gard, Michael, and Jan Wright. 2005. *The Obesity Epidemic: Science, Morality and Ideology*. New York: Routledge.

GFKBM. 2007. *Australian Better Health Initiative Diet, Exercise and Weight*. Developmental Communications Research Report. Sydney: GFK Bluemoon.

Gillick, Muriel R. 1984. "Health Promotion, jogging, and the pursuit of the moral life." *Journal of Health Politics, Policy and Law* 9 (3):369–387. doi: 10.1215/03616878-9-3-369.

Gilman, Sander. 2008. *Fat: A Cultural History of Obesity*. Cambridge: Polity Press.

Gittelsohn, A.M., J. Halpern, and R.L. Sanchez. 1991. "Income, race, and surgery in Maryland." *American Journal of Public Health* 81 (11):1435–1441. doi: 10.2105/ajph.81.11.1435.

Glassner, Barry. 1989. "Fitness and the Postmodern Self." *Journal of Health and Social Behavior* 30 (2):180–191.

Goltz, Karen, and Nina Bruni. 1995. "Health promotion discourse: language of change?" In *The Politics of Health: The Australian Experience*, edited by Heather Gardner. Melbourne: Churchill Livingstone.

Guthman, Julie. 2011a. "If they only knew: the unbearable whiteness of alternative food." In *In Cultivating Food Justice: Race, Class, and Sustainability*, edited by Alison Hope Alkon and Julian Agyeman. Cambridge, MA: MIT Press.

Guthman, Julie. 2011b. *Weighing In: Obesity, Food Justice, and the Limits of Capitalism*. Berkeley, CA: University of California Press.

Hall, Peter A. 1993. "Policy paradigms, social learning, and the state: the case of economic policymaking in Britain." *Comparative Politics* 25 (3):275–296.

Hansen, Emily, and Gary Easthope. 2007. *Lifestyle in Medicine*. New York: Routledge.

Hetzel, Basil S. 1980. *Health and Australian Society*. Third edn. Sydney: Penguin Books.

Howland, Maryhope, Jeffrey Hunger, and Traci Mann. 2012. "Friends don't let friends eat cookies: Effects of restrictive eating norms on consumption among friends." *Appetite*. 59 (2):505–509.

Irwin, A., and E. Scali. 2007. "Action on the social determinants of health: A historical perspective." *Global Public Health* 2 (3):235–256. doi: 10.1080/17441690601106304.

James, Delores. 2004. "Factors influencing food choices, dietary intake, and nutrition-related attitudes among African Americans: Application of a culturally sensitive model." *Ethnicity & Health* 9 (4):349–367. doi: 10.1080/1355785042000285375.

Jenkin, Patrick. 1981. "Economic constraints and social policy." *Social Policy & Administration* 15 (3):233–241. doi: 10.1111/j.1467-9515.1981.tb00679.x.

Krieger, Nancy. 1994. "Epidemiology and the web of causation: Has anyone seen the spider?" *Social Science & Medicine* 39 (7):887–903. doi: 10.1016/0277-9536(94)90202-x.

Krieger, Nancy. 2011. *Epidemiology and the People's Health: Theory and Context*. New York: Oxford University Press.

Leonhardt, David. 2009. "The way we live now – fat tax." *New York Times*, August 12.

Lewis, Milton J. 2003. *The People's Health: Public health in Australia, 1950 to the Present*. Westport, CT: Greenwood Press.

Lowe, Adrian. 2012. "Is this child abuse? The courts think so." *The Age*, July 12.

Lowe, M., I.H. Kerridge, and K.R. Mitchell. 1995. "'These sorts of people don't do very well': race and allocation of health care resources." *Journal of Medical Ethics* 21 (6):356–360. doi: 10.1136/jme.21.6.356.

Lupton, Deborah. 1995. *The Imperative of Health: Public Health and the Regulated Body*. London: Sage Publications.

Marmot, Michael. 2010. *Fair Society, Healthy Lives.* The Marmot Review Available from http://www.instituteofhealthequity.org/Content/FileManager/pdf/fairsocietyhealthylives.pdf.

Mayes, Christopher. 2014. "Governing through choice: Food labels and the confluence of food industry and public health to create 'healthy consumers'." *Social Theory and Health* 12 (4): 376-395.

Mayes, Christopher, and Donald B. Thompson. 2014. "Is nutritional advocacy morally indigestible? a critical analysis of the scientific and ethical implications of 'healthy' food choice discourse in liberal societies." *Public Health Ethics* 7 (2):158–169. doi: 10.1093/phe/phu013.

Murtagh, Lindsey, and David S. Ludwig. 2011. "State intervention in life-threatening childhood obesity." *JAMA* 306 (2):206–207. doi: 10.1001/jama.2011.903.

O'Brien, Paul E., Wendy A. Brown, and John B. Dixon. 2005. "Obesity, weight loss and bariatric surgery." *Medical Journal of Australia* 183 (6):310–314.

O'Hara, Lily, and Jane Gregg. 2006. "The war on obesity: a social determinant of health." *Health Promotion Journal of Australia* 17 (3) 260–263.

Oberlander, Jonathan. 2003. *The Political Life of Medicare*. Chicago, IL: University of Chicago Press.

Oldways Preservation Trust. 2013. *Claim Health and History*. [cited July 31 2013]. Available from http://oldwayspt.org/programs/african-heritage-health.

Omalu Bi, Ives D.G. Buhari A.M., et al. 2007. "Death rates and causes of death after bariatric surgery for Pennsylvania residents, 1995 to 2004." *Archives of Surgery* 142 (10):923–928. doi: 10.1001/archsurg.142.10.923.

Petersen, Alan, and Deborah Lupton. 1996. *The New Public Health: Health and Self in the Age of Risk*. St Leonards, NSW: Allen and Unwin.

Puhl, Rebecca, and Kelly D. Brownell. 2001. "Bias, discrimination, and obesity." *Obesity* 9 (12):788–805.

Rose, Geoffrey. 2001a. "Reiteration: sick individuals and sick populations." *International Journal of Epidemiology* 30:427–432.

Rose, Nikolas. 2001b. "The politics of life itself." *Theory, Culture & Society* 18 (6):1–30.

Rosen, George. 1993. *A History of Public Health*. New York: Johns Hopkins University Press.

Rothstein, W.G. 2003. *Public Health and the Risk Factor: A History of an Uneven Medical Revolution*. Rochester, NY: University of Rochester Press.

Rudolf, Mary. 2011. "Predicting babies' risk of obesity." *Archives of Disease in Childhood*. doi: 10.1136/adc.2010.197251.

Schweiker, Richard S. 1982. "Strategies for disease prevention and health promotion in the Department of Health and Human Services." *Public Health Report* 97 (3):196–198.

Shute, Nancy. 2012. *Cooking Up Change: How Food Helped Fuel the Civil Rights Movement*. [cited July 31 2013]. Available from http://www.npr.org/blogs/thesalt/2012/01/16/145179885/cooking-up-change-how-food-helped-fuel-the-civil-rights-movement.

Skrabanek, Petr. 1994. *The Death of Humane Medicine and the Rise of Coercive Healthism*. London: Social Affairs Unit.

Sorlie, P., E. Rogot, R. Anderson, N.J. Johnson, and E. Backlund. 1992. "Black–white mortality differences by family income." *The Lancet* 340 (8815):346–350. doi: 10.1016/0140-6736(92)91413-3.

Terris, Milton. 1992. "Healthy lifestyles: the perspective of epidemiology." *Journal of Public Health Policy* 13 (2):186–194.

Terris, Milton. 1999. "The neoliberal triad of anti-health reforms: government budget cutting, deregulation, and privatization." *Journal of Public Health Policy* 20 (2):149–167.

Tesh, Sylvia N. 1988. *Hidden Arguments: Political Ideology and Disease Prevention Policy*: Rutgers University Press.

Wain, Harry. 1970. *A History of Preventive Medicine*. Springfield, IL: Charles C Thomas.

Whitehead, T. L. (1992). "In search of soul food and meaning: culture, food, and health". In H. A. Baer and Y. Jones (eds), *African Americans in the South: Issues of Race, Class, and Gender*. Athens, GA: University of Georgia Press.

World Health Organization. 1986. *Ottawa Charter for Health Promotion*. [cited January 16 2012]. Available from http://www.who.int/hpr/NPH/docs/ottawa_charter_hp.pdf.

Wyklicky, Helmut, and Manfred Skopec. 1983. "Ignaz Philipp Semmelweis, the prophet of bacteriology." *Infection Control* 4 (5):367–370.

Žižek, Slavoj. 2009. *Violence: Six Sideways Reflections*. London: Profile Books.

# 4 Lifestyle as identity

## Consumption and the ethics of the self

> But couldn't everyone's life become a work of art? Why should the lamp or the house be an art object, but not our life?
>
> Foucault, *On the Genealogy of Ethics* (1983: 236)

Neoliberal ideas of governing and epidemiological knowledge form a grid of intelligibility in which individual choices are made visible in relation to obesity, health and security. The visibility of these choices enables not only strategies of governance, but also practices of the self. The enabling network of lifestyle is comprised of lines of subjectification as well as lines of knowledge and power. As such, it is not just governments or 'the State' interested in seeing choices and lifestyle, but the individual and her peers. Sociological analyses of lifestyle in consumer societies offer a significant addition to governmentality perspectives on the process of subjectification.

The sociological notion of lifestyle as the conscious styling of life through consumer choice and leisure activity pre-dates the epidemiological and public health concepts of lifestyle as behaviours associated with chronic diseases. From the advent of mass-produced consumer goods in the late nineteenth century, sociological research has examined the significance of everyday objects and products in *styling* life, affirming identities and differentiating social groups. Shopping, recreation, hobbies, or culinary preferences serve as opportunities for individuals to make choices that reflect and reinforce individuality and belonging. This sociological research is essential for articulating the active role of the individual in forming a healthy identity.

The sociology of lifestyle is a rich and diverse literature. As such, breadth will need to make way for depth. My focus is particularly on Pierre Bourdieu's examination of lifestyle in France from the mid to late 1960s, as well as the work of Henri Lefebvre and Guy Debord on consumerism. Although there are significant differences among these sociologists, I draw on their work to enrich and deepen Foucault's outline of the ethics of the self. The ethics of the self provides useful clues for considering subject formation in everyday life and the consumer practices in the lifestyle network. However, there is a rub between sociological analyses and Foucault's approach. Addressing the tension between sociological analyses of lifestyles as consumer-seduction and Foucauldian analyses of biopolitical governance through consumption (Binkley 2007b: 120), I suggest that a strategic logic defines the relationship between neoliberal 'rationalization of everyday life' and the commodity seduction described by sociology (Larsen 2011: 221). Conceiving the relation as strategic or agonistic explains the governmental rationality that operates through and with the active engagement of the subject. As such, the struggle between the stylization of life and the securitization of life produces a subject.

## The seduction and styling of everyday objects

Sociology has a long tradition in analysing lifestyle as related to consumer choice, identity and social status. Lifestyle entered the sociological lexicon through the translation of Max Weber's three terms *Lebensführung* (life conduct), *Lebensstil* (lifestyle) and *Lebenschancen* (life chances) as 'lifestyle' (Abel and Cockerham 1993). This reduction of three terms into a single concept is problematic, especially due to the eradication of life chances from the English concept. Weber's social and economic analysis emphasized the role of life chances as the conditions that structure and make possible certain life styles and conducts. As Abel and Cockerham note, 'lifestyles are based on choices (*Lebensführung*), but these choices are dependent upon the individual's potential (*Lebenschancen*) for realizing them' (1993: 554). The simplified notion of lifestyle, as used in much sociological, political and public health, emphasizes individual choice without giving adequate attention to the life chances that precede such choices.

Despite debates surrounding translations of Weber, lifestyle has become widespread in Anglo-American sociology to conceptualize 'alternative ways of living, usually conspicuous through values and modes of consumption, which are attendant upon the increasing differentiation of advanced capitalist societies' (*Concise Oxford Dictionary of Sociology* 1994). The use of lifestyle in sociology is polyvalent. Mike Featherstone defines lifestyle as:

> individuality, self-expression, and a stylistic self-consciousness. One's body, clothes, leisure pastimes, eating and drinking preferences, home, car, choice of holidays etc. are regarded as indicators of the individuality of taste and sense of style of the owner/consumer.
>
> (Featherstone 1987: 55)

For Anthony Giddens, lifestyle is 'routinised practices, the routines incorporated into habits of dress, eating, modes of acting and favoured milieux for encountering others' (1991: 81). Annamarie Jagose suggests that, 'lifestyle draws promiscuously on a range of concepts such as taste, income, health status, diet, aspiration, subculture and leisure in order to represent everyday life in advanced capitalist cultures' (2003: 109). These definitions do not exhaust but represent the broader sociological literature on lifestyle, and point toward the role of lifestyle in structuring everyday activities through routine, habit and self-expression.

Sociological analyses of lifestyle generally address two broad themes: Marxist ideas of alienation, and Weberian notions of social structure. The alienation motif suggests that the consuming individual is not only alienated as a worker from the products she produces, but she is further alienated by the seduction of commodities that do not fulfil the purported promises of deepening, enriching or actualizing the self. This type of analysis is common in the popular critiques of consumer society put forward by Naomi Klein, Kalle Lasn and others (Klein 2002, Sharpe 2003, Lasn 1999). On this view the subject is hard-wired to be free, yet corporations manipulate and estrange individuals from true humanity through false and deceptive advertising. Thus the goal of Klein, Lasn and anti-globalization activists is to 'jam' the transmission of corporate advertising to enable individuals to have authentic social relations.

In addition to alienation, Weber's 'iron cage' influences much of the analysis of lifestyle. Weber famously wrote, 'In Baxter's view the care for external goods should only lie on the shoulders of the "saint like a light cloak, which can be thrown aside at any moment." But fate decreed that the cloak should become an iron cage' (Weber 2001: 123). There has been debate over this translation (Baehr 2001), however the point is that the goods produced and consumed under economic and structural conditions ensnare and construct the subject in a steel case or cage. Weber's analysis of the cage has led to critiques that he pays too much attention to the objective or structural influences that determine subjects and does not leave room for the subject or agent to act. The structure/agent tension, akin to life-chance/life-choice, is implicit in much of the sociological analyses of lifestyle and consumer society.

In contrast to Marxist and Weberian analyses of lifestyle, a Foucauldian analysis is not primarily concerned with alienation. This is particularly the case in the context of neoliberalism. The neoliberal subject (*homo œconomicus*) is an empty subject that is filled and produced through practices of consumption. There is not a true or deep subject to be liberated or reconciled. Below I argue that agonism is an appropriate way to understand the dissensus between individual consumption and objectives of (bio)political rationalities. Bearing the underlying influence of themes of alienation and social structure in mind, I outline several themes in greater detail that are pertinent to understanding the lifestyle *dispositif* that produces its subject. These include the conspicuousness and spectacle of everyday objects, the structuring of identities, and signification of group belonging.

An early figure in the analysis of the everyday objects in twentieth-century consumer societies was Henri Lefebvre. Lefebvre's three-volume *Critique of Everyday Life* (1947 to 1981) is a central text in the sociology of consumption. Working primarily within a Marxist perspective, though with Nietzschean gestures, Lefebvre sought to redress what he believed to be the sociological neglect of the concept of alienation (Lefebvre 1996: 3). While the alienation of the worker is extensively analysed, Lefebvre argued that greater focus was required on the commodification of everyday life that results in alienation in the home, leisure and holiday time in Western societies (Lefebvre 1996: 76).

Lefebvre's work draws attention to everyday life as a field of social and philosophical significance in producing the subject. The influx of modern consumer goods following the wars of the first half of the twentieth century influenced relations among individuals, objects and work in Western societies. Fridges, televisions and washing machines were enthusiastically embraced for their efficiency, but also their aesthetic and symbolic appeal. Consumer products and 'everyday objects suddenly shone with the transformative power of the sublime' (Shields 1999: 66). Writing in 1958 for the tenth anniversary of the first volume of *Critique of Everyday Life*, Lefebvre observes:

> Problems of everyday life and studies of everyday life have become increasingly important in the minds of historians, ethnographers, philosophers, sociologists...Our very best informed and most 'modern' publications – daily and weekly newspapers, reviews – have started columns dealing with everyday life.
>
> (Lefebvre 1996: 7)

Lefebvre proposed that the entire structure of human interactions could be understood if greater attention was paid to the 'atomic structures' (Merrifield 2006: 5). Rather than understanding society through grand theories, the minute, mundane and background provide the avenues through which social and political interactions are understood. Commonly regarded as 'residual, defined by "what is left over"' (Lefebvre 1996: 97), Lefebvre contends that everyday life:

> is profoundly related to *all* activities, and encompasses them with all their differences and their conflicts; it is their meeting place, their bond, their common ground. And it is in everyday life that the sum total of relations which make the human – and every human being – a whole takes its shape and form.
>
> (Lefebvre 1996: 97)

Everyday objects and daily habits are not the remnant once specialized, structured or superior activities are extracted, but the space that connects all activity. In contrast to metaphors of 'mountain tops' used to describe creative moments, Lefebvre maintains, everyday life is a 'fertile soil' (Lefebvre 1996: 87). The flowers and vegetation may capture the eye, but it is 'the earth beneath, which has a secret life and a richness of its own' (Lefebvre 1996: 87). Although his poeticism may suggest a romantic valorization of the quotidian, Lefebvre was more interested in emphasizing the inescapable connectedness between everyday life and the spectacular: 'the ordinary is epic just as the epic is ordinary' (Merrifield 2006: 7).

The sociological work of Lefebvre on the significance of mid-century consumption in Europe continues to influence the examination of everyday objects and consumer products in subject formation (de Certeau 1988, Elden 2004). Lefebvre's work is also significant for the lines of connection established between mundane objects and the political, economic and social. These influences and cues established by Lefebvre are further elaborated and articulated in the work of his student and colleague Guy Debord.

In *The Society of the Spectacle*, Debord introduces the idea of the everyday as spectacle mediated through images and commodities. Debord begins by writing, 'all of life presents itself as an immense accumulation of spectacles. Everything that was directly lived has moved away into a representation' (Debord 1983: 1). The representation of life through spectacles is an idea that has grown in significance with the pervasiveness of television, and especially with the Internet, mobile devices and social networking. Visual mediums of instantaneous communication have led to the ubiquity of the image as a representation of life. Like Foucault's relations of power, the spectacle is the relation among people, not the images themselves. 'The spectacle is not a collection of images' writes Debord, 'but a social relation among people, mediated by images' (Debord 1983: 4 ). Debord's analysis demonstrates the increased visibility and significance of the everyday through advertising, images and television.[1]

Like Lefebvre, Debord is concerned with alienation, arguing that the commodity occupies social life through representing life back to 'the fragmented individual' (Debord 1983: 42). My interest in Lefebvre's and Debord's analyses of the everyday is not alienation but the biopolitical governance via everyday consumption that produces or forms the self. Debord's concept of the spectacle can be usefully employed to further examine the operation of biopolitics in the interaction between commodity, the everyday, and the subject. Sadie Plant draws on Debord to describe individuals as '[b]ombarded by images and commodities which effectively represent their lives to them…[and] experience reality as second-hand' (1995: 10). The spectacle of the everyday in consumer society alters the social relations between individuals and produces new modes of subjectification. This is particularly evident where the body and consumer choice come to represent health, beauty and responsibility. The bodies and choices of individuals are enticed into 'self-betterment strategies and technologies' that have become entwined with desires for personal fulfilment, such that the 'internal search for self is becoming a driving force for physical activity' (Howell and Ingham 2001: 343). Ideas of self-improvement and investment are powerfully mobilized in neoliberal governmental strategies that encourage individuals to adopt a healthy lifestyle (Lazzarato 2009, McNay 2009).

While there are aspects of Lefebvre's and Debord's work that conflict with Foucault's, the focus on everyday life and the spectacle of life represented via images can be used in concert with Foucault's notion of surveillance to assist in developing a critical response to the obesity epidemic. An attempt to dovetail the spectacle with surveillance may appear careless in light of Foucault's apparent dismissal of such a merger in *Discipline and Punish*. Writing shortly after the publication of *The Society of the Spectacle*, Foucault states, 'our society is one not of spectacle, but of surveillance' (Foucault 1991: 217). According to Foucault, it is through the mechanism of the all-seeing panopticon that individuals are fabricated and invested with the effects of power, not the mediation of images. Foucault emphasizes his rejection of the spectacle in stating, 'we are neither in the amphitheatre, nor on the stage but in the panoptic machine' (Foucault 1991: 217). However, perhaps Foucault's dismissal of the stage and the spectacle was too hasty.

Jonathan Crary speculates that Foucault did not spend enough time watching television to appreciate Debord's spectacle, suggesting that in television '*surveillance* and *spectacle* are not opposed terms…but collapsed onto one another in a more effective disciplinary apparatus' (Crary 1989: 105). The dovetailing or collapsing of surveillance and spectacle has further intensified with the spread of social media and mobile devices. Similar to Crary, Debord recognizes an important relationship between surveillance and spectacle. In *Comments on the Society of the Spectacle* Debord states that 'from the networks of promotion/control one slides imperceptibly into networks of surveillance/disinformation' (Debord 1998: 74). Rather than conflicting with Foucault's notion of surveillance, the spectacle offers new modes of surveillance through which individuals modify, shape and style their life. Through the colonization of the everyday as spectacle by images of advertising mediated throughout society, the individual's everyday consumption and body are made conspicuous and more visible as the object of surveillance. Not only are consumptive practices made conspicuous attracting both admiration and surveillance, but the diverse array of everyday objects and choices are fragments that are structured by lifestyle into a coherent and readable subjectivity.

The merger of spectacle and surveillance is seen in the *Measure Up* social marketing campaign. In surveying the spectacle of the morphing body, the viewer is encouraged to discipline their own body and choices to avoid a similar future. Through the 'stage' of television, Internet or smartphones, the healthy body is a visual commodity signifying 'lifestyle success and lifestyle mobility' (Howell and Ingham 2001: 336). The body is positioned as an object of surveillance and spectacle that communicates one's ability to form a self that styles and secures life.

## Lifestyle, identity and group belonging

Prominent figures in sociological analyses of lifestyle in the late twentieth century – Anthony Giddens, Mike Featherstone and Pierre Bourdieu – suggest that the decline of traditional sources of identity formation led to a valorization of consumer practices. Categories of class, religious adherence or ethnicity did not disappear, but became fragmented and less determinative. In their place emerged new forms of self-creation and the stylization of life that revolved around consumer practices and suburban living. Sociologists, as well as advertisers, turned to the idea of lifestyle to articulate the processes of structuring disparate consumer and leisure practices to form specific subjectivities. Advances in production techniques allowed for choice and a plurality of forms of life that emphasized culture and aesthetics over economics and function. Everyday objects came to represent and signify symbolic power and status to those possessing them.[2] In this context, Giddens contends that lifestyle serves as an ordering mechanism that governs the individual's interaction with the everyday and enables the formation of subjectivities that are distinguished from and bound to social groups (1991: 2).

In *Modernity and Self-Identity* Giddens is interested in the 'emergence of new mechanisms of self-identity' that shape, and are shaped by, the institutions of modernity. These mechanisms operate in the contemporary situation, which Giddens describes as 'a post-traditional order' where 'the sureties of tradition and habit have been replaced by the certitude of rational knowledge'. This certitude, however, is mediated through experts who guide individuals towards particular lifestyle choices. The self becomes a reflexive project that individuals are engaged in with expert guidance. However, self-as-project is never fixed or resolved but in a continual process of flux and becoming due to the continual imperative to choose (1991: 70ff).

The uncertain and multiple choices in life without tradition or habit have given birth to new experts and authorities (Mayes and Thompson 2014). Dieticians, personal trainers, financial planners, life coaches and medical professionals reinforce and respond to the climate of doubt and uncertainty seeking to establish 'ontological security' for individuals. According to Giddens, these experts and mechanisms of self-identity provide 'a "protective cocoon" which stands guard over the self in its dealings with everyday reality' (1991: 3). Lifestyle as a 'mechanism of self-identity' offers 'stability' or 'security' in the form of navigating the instability and uncertainty of the everyday (Binkley 2007a: 185ff). Giddens argues that in the post-traditional order, 'the self, like the broader institutional contexts in which it exists, has to be reflexively made…amid a puzzling diversity of options and possibilities' (1991: 3). Lifestyle becomes a significant mechanism of identity formation and group belonging within this situation of disorienting choice and open possibilities.

Owing to the increase of choice, Giddens suggests that we are now faced with a situation where we 'not only follow lifestyles, but in an important sense are forced to do so – we have no choice but to choose' (1991: 81). Whether we opt out or in, we are faced with the choice of choosing a particular lifestyle as an 'integrated set of practices which an individual embraces…because they give material form to a particular narrative of self-identity' (Giddens 1991: 81). Even the decision to remove oneself completely from consumer choices becomes interpreted as an 'alternative' or 'off-the-grid' lifestyle.

In addition to forming one's own identity through engagement or disengagement from such choices, lifestyle serves to represent the self as belonging to a group or class. Lifestyle as distinction and group belonging is profoundly articulated in the work of Pierre Bourdieu. Bourdieu defines lifestyle as 'a system of classified and classifying practices, i.e. distinctive signs ("tastes")' (1984: 171). The tastes or preferences of individuals expressed in consumption and bodily deportment serve to transmute 'things into distinct and distinctive signs' (Bourdieu 1984: 174). These signs give the subject distinction and indicate group or class membership. Like Foucault, Bourdieu developed and used a number of original analytic tools and concepts. *Field, capital* and *habitus* are of particular relevance for my argument and are central to Bourdieu's formula: [(habitus) (capital)] + field = practice (1984: 101).

In some ways Bourdieu's habitus is akin to Foucault's *dispositif*. It makes practices, activities and identities intelligible and visible. Habitus is a 'structuring structure' and a 'structured structure' (Bourdieu 1984: 170). Not only does the habitus structure or give form to the practices of individuals into a social identity but it is structured by the social field in which it is situated. The habitus sits within a field, which is a dynamic and temporal space that produces and regulates the habitus and enables the recognition of capital as economic or symbolic (Moore 2008: 105). A dialectic logic operates between the field and the habitus, which is Bourdieu's attempt to get beyond the binary of structure and agency that has occupied sociological analyses since Durkheim and Weber.

The field and habitus do not completely overlap, but form a dialectic. According to Thompson, it is through the dialectic of field and habitus that 'specific practices produce and reproduce the social world that at the same time is making them' (Thomson 2008: 75). The subject or agent engaged in practices makes and remakes the social world in which they are situated. The relation between the subject and the social world can be smooth when there is a close overlap between the habitus and field. Robert Moore describes this as being a 'fish in water' moment. The subject is attenuated to the unwritten rules of the social world and can navigate them. However, when the subject is disconnected or feels anxious, this is a 'fish out of water' moment where the habitus and field are clashing (Moore 2008: 57).

Habitus is defined by the relationship between 'the capacity to produce classifiable practices' and the 'capacity to differentiate and appreciate these practices and products (taste)' (Bourdieu 1984: 170). It is in this social space of habitus that Bourdieu's notion of lifestyle is constituted. By generating practices that classify those engaging in them, habitus enables the possibility for individuals to distinguish and select different ways to engage with classifying practices. Individuals with taste, in the normative sense of being able to discern and judge, are able to navigate the classifying practices to select those that are distinctive of a particular disposition and thereby establish a social identity or lifestyle (Bourdieu 1984: 172).

Bourdieu's notion of capital is essential to an understanding of the way lifestyles signal social identity. There are a number of sub-genres of capital, however there is a primary distinction between economic and symbolic capital (Moore 2008: 103). Economic capital purports to allow transparent exchange to achieve instrumental ends. For example, Student A may pursue a tertiary degree for the purpose of attaining a high paying job in order to purchase material goods. The degree in this context has instrumental value and the transaction is relatively transparent. Symbolic capital, on the other hand, suppresses the idea of instrumental ends and instead claims to hold intrinsic worth. Student B may purse the same tertiary degree but claim this is purely for the attainment of knowledge as an end in itself. In this context, the degree is purported to have intrinsic (symbolic) value that cannot be reduced to an economic transaction. Bourdieu however, considers the value of symbolical capital to not be intrinsic but to be representing the assets that confer advantage or disadvantage in different social contexts.

Despite claims to the contrary, symbolic capital has instrumental value that is established through a transaction process governed by codes of tastes and distinctions. A primary field where this transaction occurs is consumerism. Bourdieu considers practices of consumption as 'a process of communication, that is, an act of deciphering, which presupposes practical or explicit mastery of a cipher or code' (Bourdieu 1984: 2). To decipher the code an individual needs knowledge of taste or distinction. For example, two different shoppers enter the habitus of a vinyl record store in SoHo. Shopper A has taste and is at home in the habitus. She understands the ordering of the physical space. She knows what to expect of the staff and how to talk with them. She recognizes the music that is playing over the stereo and the posters on the walls. She is disposed to the classifying practices and has the taste to distinguish them. Approaching the counter she says 'Hey, I was in here last week and I was talking to Dave. He said you were getting some *Hüsker Dü* in. I am keen to get my hands on *Zen Arcade*. Has anything arrived?' Shopper B however is trying to find a gift for a friend. He doesn't really listen to much music, except what is on commercial radio. He enters the store. The music is disorienting. He looks to the shelves but doesn't recognize any names or understand the system of ordering. Feeling uneasy, he waits nervously near a staff member engaged in an animated conversation with another customer about something he doesn't understand. He waits. They look at him and continue talking. There is a pause and he quickly asks 'I'm looking for a gift…for a friend. I am not quite sure though. I don't know if he has a record player. Do you have CDs or iTunes vouchers?'

The taste of the shoppers classifies them as part of an at-home in the habitus or as outside it. Bourdieu writes that social subjects are 'classified by their classifications [and] distinguish themselves by the distinctions they make, between beautiful and the ugly, the distinguished and the vulgar, in which their position in the objective classification is expressed or betrayed' (Bourdieu 1984: 6). In the habitus of the vinyl store, Shopper A could be described as a beautiful subject, a subject with taste that belongs in the habitus. Shopper B, however, has vulgar taste, does not belong, and is classified accordingly.

The capacity of taste is symbolic capital that is the 'generative formula of lifestyle' serving to structure 'a unitary set of distinctive preferences which express the same expressive intention in the specific logic of each of the symbolic sub-spaces, furniture, clothing, language or body hexis' (Bourdieu 1984: 173). For an individual to tastefully navigate a habitus it is not only a matter of buying the right record at the right store, but choreographing multiple practices into a distinctive whole. Engaging with practices, such as artistic, culinary or educational activities, individuals are distinguished as part of a group or class (Bourdieu 1984: 176). Much more can be drawn from Bourdieu's analysis of lifestyle, habitus and capital. For the present purpose the primary point is that lifestyles are 'the systematic products of habitus, which, perceived in their mutual relations through the schemes of the habitus, become sign systems that are socially qualified' as ugly or beautiful, fit or fat, healthy or unhealthy (Bourdieu 1984: 172).

Despite their friendship, shared influences and similar research interests, Foucault did not explicitly engage with Bourdieu's work on subject formation. The reasons for this non-engagement are not clear, however David Hoy suggests that the sociology of Bourdieu, 'can be read as deepening Foucault's account of how subjectivity is constructed through power relations by providing a much more detailed sociological theory of this process' (Hoy 1999: 11). With the discussion of Bourdieu in mind, I turn to Foucault's analyses of subject formation.

## Agonistic consumption: everyday practices and subject formation

In 1982, Foucault gave a seminar at the University of Vermont critically reflecting on the relationship between his earlier work on power and discipline and his interest in practices of the self in the ancient world (Foucault 2000b). Foucault acknowledged that technologies of power and domination had occupied much of his attention, to the neglect of technologies of the self. Foucault's turn to Greek and Christian practices of the self in *The Use of Pleasure* and *The Care of the Self* is part of his interest in governmentality. Governmentality is the combination of technologies of domination and technologies of the self. Although ancient Greek practices of the self are far removed from the present in which everyday life is rationalized, consumed and invested, Foucault's work on the ethics of the self and subject formation provides useful indicators to address the creation of a 'healthy lifestyle' in neoliberal societies (O'Leary 2003). Further, the interaction between technologies of power and technologies of the self is central to my argument that lifestyle is a biopolitical network that styles the individual through consumer practices to accord with norms of health that secure the population.

The aesthetic practices of consumption merge with the biopolitical norms of health to produce a subject. This merger, however, is not final or complete. The relationship between technologies of power and technologies of the self is ongoing and agonistic. By describing the relationship as agonistic, I contend that the subject is constituted through struggles over consumption. These struggles are not resolved through rational consensus but are continually held in tension. For example, there are different and competing reasons for eating food: health, pleasure, custom, ritual, affordability, availability, and so on. A rationalist approach promoted via public health messages is something along the lines of whatever promotes health and prevents disease, yet this is rarely the way people encounter food or dieting over the long term.

The self is formed through agonistic struggle between different practices and choices that have competing underlying logics. These struggles occur between technologies of the self that use consumer practices to form the self-as-art and technologies of power that use consumer practices to form the self-as-responsible. While the first styles life through commodity seduction, the second secures life by encouraging responsible choice. Not only does analysing consumption as agonistic serve to articulate the process of self-formation in the lifestyle network, but it also highlights a distinction between a governmentality perspective and sociology of lifestyle. Although governmentality studies have successfully described strategies of governance and relations of power, they have been accused of giving inadequate accounts of the individual in practices of consumption (Binkley 2007b). In contrast, sociological analyses have been successful in describing self-formation through consumer lifestyle practices;

yet have inadequately addressed the influence of politics and technologies of domination (Rose 1999).

By drawing together sociological and governmentality perspectives on consumption and lifestyle, I contend that an agonistic logic, what Foucault has called strategic logic, operates in subject formation (2008: 42). While the dialectic logic of Bourdieu attempts to homogenize contradictory terms and practices by aligning the subject, habitus and field (Moore 2008: 57, Thomson 2008: 75), the agonistic logic holds heterogeneous practices together. The dialectic logic moves from seduction towards responsibility, while an agonistic logic is both seduction and responsibility, style and security. Apposite yet trivial examples include sugar-free soft drink, carb-free beer and chocolate purporting to reduce the risk of type 2 diabetes (Wedick et al. 2012, BBC News 2008). It is not a logic of either/or but of both/and. Similarly, Slavoj Žižek uses the example of chocolate laxatives to describe the both/and logic of 'liberal-communists', such as Bill Gates and George Soros, who 'give away with one hand what they first took with the other' (Žižek 2009: 18). However, before continuing with this discussion, I need to outline Foucault's analysis of technologies of the self in the context of Greek subject formation.

In his final years, Foucault directed his attention toward Greco-Roman ethics and the constitution of the ethical subject around the use of pleasure. In the monographs *The Use of Pleasure* and *The Care of the Self*, as well as lectures, seminars and interviews, Foucault investigates the constitution of the subject through specific techniques, practices and knowledges of the self. Foucault describes this project as a 'genealogy of the subject as a subject of ethical actions' (Foucault 1983: 240). Ethics in this context does not indicate principle based normative action that if followed is considered morally praiseworthy. Ethics indicates what Foucault variously calls the 'art of life', 'structure of existence', 'practice of freedom' or the 'relation to oneself' (Foucault 1983). Ethics in this sense is 'a way of being and of behaviour…[as] a mode of being for the subject, along with a certain way of acting, a way visible to others' (Foucault 2000a: 286). Ethics is a practice that forms the subject, guides action, and mediates the subject's relation to oneself, others and the world. Given this, 'ethics', '*ēthos*', 'way of life', and 'lifestyle' form a kinship or network of concepts that are neither synonymous nor entirely distinct.[3] This construal of ethics has two important implications. The first is that ethical practice occurs in everyday and mundane activity. A person's *ēthos* is demonstrated by clothing and style as well as embodied in gait and manner in which one responds to an event (Foucault 2000a: 286). In this way the free Greek citizen demonstrates a particular *ēthos* or art of living by the way they dress, mourn, walk or eat (Hadot 1995a, b: 206ff). The second implication is that through such activities the subject is brought into existence into the social world. Ethics is made visible to others and it is also a practice that one performs on oneself.

In the *Use of Pleasure* Foucault initiates his new direction of enquiry into technologies of the self by asking: 'How did sexual behaviour…come to be conceived as a domain of moral experience?' (Foucault 1992: 24) Through philosophical-historical analysis, Foucault shows why particular sexual activities of the free Greek male in specific contexts became a point of anxiety and debate. In seeking to answer this question Foucault moves toward identifying modes of subjectification and the analysis of the 'forms and modalities of the relation to self by which the individual constitutes and recognizes himself *qua* subject' (Foucault 1992: 6).

In this analysis of the constitution of the subject through an ethics understood as practice, Foucault isolates four aspects in the process of subjectification: the ethical substance, the mode of subjection, the ascetic work, and the *telos* of the ethical subject. The relationship among the four aspects is not linear but dynamic and interconnected. Foucault writes that '[t]here are both relationships between them and a certain kind of independence' (1983: 240). Foucault describes the four aspects in the Greek context, suggesting that the 'ethical substance' was *aphrodisia* (pleasure), the 'mode of subjectification' was politico-aesthetic choice, the 'ascetic form' or 'ethical work' was a variety of the techniques used to govern particular relations around *aphrodisia*, and finally the 'telos' of subjectification was the mastery of oneself (Foucault 1983: 241).[4]

The four aspects of subject formation serve to shape and style 'how the individual is supposed to constitute himself as a moral subject of his own actions' (Foucault 1983: 238), where the operation of the four aspects is neither linear nor completely open, but works through a variety of everyday practices of the individual. A crucial aspect of the process of subject formation is the use of techniques, mechanisms and strategies to work on and fashion the self – the ascetic aspect. The 'ethical work', or 'ascetic work', can take a variety of forms in order 'to bring one's conduct into compliance with a given rule [and] to attempt to transform oneself into the ethical subject of one's behaviour' (Foucault 1992: 27). The 'ethical work' employs techniques that enable and guide one's conduct to conformity with the 'rule of conduct'. It is important to note that in this context 'ethical work' do not simply refer to austere practices of renunciation or deprivation, rather Foucault refers to it as also 'the means by which we can change ourselves in order to become ethical subjects' (1983: 239). The ascetic aspect is the way in which one works on oneself in relation to the mode of subjection, the techniques of 'the self-forming activity' (Foucault 1983: 239). Examples include writing in notebooks, memory exercises, meditating on one's death, or abstaining or engaging in certain physical activities at specific times.

The technologies and practices through which the subject is constituted demonstrate the interaction between Foucault's work on subjection (self/ aesthetics) and power (production/domination). While historically removed from neoliberal and consumer societies, Foucault's analysis of Greek subject formation indicates how the ethical subject can be formed around everyday practices in contemporary neoliberal societies that mobilize a notion of lifestyle to govern obesity.

Ethical work in the current context occurs through a variety of avenues. The ability of the subject to choose 'healthily' forms indicates the subject's ability to negotiate and use free choice. The ethical work is not equivalent to frugality or renunciation – there is a strong imperative to continue consuming – but to consume in accordance with norms of health leading to security. There are a variety of techniques, knowledges and experts that assist the individual in becoming a healthy subject through everyday choices. For instance, the *Let's Move* and *Change4Life* health campaigns produce menus and exercise planners to assist in the healthy consumption of food (Department of Health 2011, Let's Move 2011). Front-of-pack food labelling informs consumers about the nutritional context of packaged foods (Mayes 2014). Celebrities like Jamie Oliver and Michelle Obama offer advice through a variety of media. And the Internet and social media provide opportunities to document choices and seek advice from peers. These avenues circulate the norms of health that make it possible to recognize, measure, and assess the success or failure of the subject to embody a healthy lifestyle. These locales of lifestyle guidance position bodies and associate activities as 'the medium through which we engage the world and are recognized by others as subjects' (Heyes 2007: 17).

The subjectification process in the lifestyle network is not enforced by a compulsory set of requirements. In contrast to the disciplinary society and technologies of power analysed in *Discipline and Punish* and *The Will to Knowledge*, the ethics and technologies of the self examined in *The Use of Pleasure* and *The Care of the Self* are more about self-stylization and choice than about population normalization and domination. However, a tension remains between power relations directed towards security and ethics of the self that stylize life. This tension is evident in the overlap between lifestyle epidemiology, which medicalizes everyday practices and choices, and lifestyle consumption, which aestheticizes everyday practices and choices. However, the *dispositif* seeks to hold these heterogeneous technologies and knowledges in connection in order to constitute the biopolitical subject. The biopolitical emphasis on fostering life employs 'norms of discipline' and 'norms of regulation' that create an imperative for the subject to make healthy choices in everyday life in order to promote health and secure the future. Yet, these choices and practices are also aestheticized and transformed such that the healthy choice is the beautiful choice. In following chapter I use examples from government health campaigns, lifestyle media, smartphone apps and medical discourse to demonstrate the way the lifestyle *dispositif* blurs the lines between consumerism and health.

Importantly, the agonism of choice within the *dispositif* cannot be deferred. Making a similar observation to Giddens, Nikolas Rose states that 'modern individuals are not merely "free to choose", but *obliged to be free*, to understand and enact their lives in terms of choice' (1999: 87). The biopolitical rationality operating through the lifestyle network uses norms of health and the body to direct individuals to make responsible and health promoting choices. In tension with this is the logic of consumer seduction described by sociology of lifestyle that entices consumers to freely choose and indulge desires. Lifestyle indicates and structures the individual's relationship to the practices of freedom and in so doing reveals the politics of everyday life. The use of freedom conveyed by lifestyle is demonstrated through economic, physiological and aesthetic markers read by the self and others. In the context of lifestyle epidemiology, neoliberal health policies and consumerism, the everyday choice of individuals indicates the success or failure of the subject to form a healthy lifestyle.

While the individual is ultimately responsible for choice, there is an increasing array of experts and authorities to guide choice in everyday activity. Traditional experts such as priests, physicians, philosophers or scientists have multiplied and splintered into life coaches, nutritionists, personal trainers, financial planners, interior designers and so on. These experts guide and compound the agonism of consumer choice by lending authoritative and scientific knowledge. The navigation of consumer choice in tension with expert knowledge shapes and constitutes the subject. Furthermore, with the withdrawal of the State and greater emphasis placed on the entrepreneurial individual, there is a requirement to exercise 'smart lifestyle choices' that 'provide personal freedom and opportunity to share in the good life: To control one's own future' (Howell and Ingham 2001: 41). Intimate relations can be formed with expert guides but success or failure in achieving a healthy lifestyle lies with the individual.

The sociological literature on lifestyle described here serves as a helpful corrective to biopolitical perspectives that focus on population level strategies to the neglect of the individual. The individual is not passive but actively adopts, chooses and creates a lifestyle. According to Featherstone, lifestyle enables the individual to actively style life through 'playful exploration of transitory experiences and surface aesthetic effects' (1991: 95). With the abundance of choice, lifestyle is a mechanism of playful identity formation. Jagose also argues that in post-traditional consumer society 'lifestyles might usefully be considered as tactical compensation for such an absence or breakdown, as ways in which the material artifacts of mass culture are used to articulate an individual's identity' (2003: 111). In this sense, lifestyle is an aesthetic practice of the self, which considers consumer practices and everyday activities as opportunities to cultivate the self into a work of art to be viewed by others.

Sociological descriptions and analyses of lifestyle may provide a more in-depth account of the individual and consumer practices than is ordinarily found in the governmentality literature. However, these analyses have failed to adequately account for biopolitical realities of the lifestyle as an enabling network that responds to the urgent need of obesity and activates technologies of power and the self. The formation of the self through lifestyle and consumption has been characterized as resembling the 'commodities themselves, high on "bling", yet lacking in depth, durability, and lasting meaning' (Binkley 2007b: 115). Zygmunt Bauman argues that individual freedom and consumer choice in late capitalist societies have led to vacuous subjectivities (2000: 35). And Thomas Lemke writes that Giddens's discussion of lifestyle and life-politics is 'curiously apolitical because it completely lacks the subversive and resistant moments that transcend the modern order' (2011: 85–86). Foucault's ethics of the self has also been subject to a similar critique. Lois McNay contends that Foucault's aesthetics of the self focuses too heavily on the 'stylization of daily life' and 'amounts to an amoral project for privileged minorities' (1994: 133–134).

In attempting to correct his overemphasis on the technologies of domination by including technologies of the self, Foucault is now exposed to criticisms of offering an apolitical aesthetics of life. Likewise, the sociological analysis of consumer lifestyles offers a rich basis for understanding subject formation through individual practices, yet it does not give sufficient attention to political forces. My analysis of lifestyle as an enabling network understands consumption as an agonistic activity that holds heterogeneous practices together to form the self. In this network, technologies of the self struggle with technologies of domination to constitute biopolitical subjects that are responsible and entrepreneurial. Not all individuals will be able or will want to participate in this agonistic struggle. In some senses McNay and Lemke are correct in stating that the aesthetics of the self are for a minority and lifestyle does not offer tools of resistance or subversion. The aesthetic orientation of lifestyle may be apolitical and vacuous for a minority. However, the focus on resistance and political lifestyle is not on the 'alienation' of a majority who think a healthy lifestyle is authentic existence, but those who do not or cannot conform to that minority and are exposed as 'ugly', 'pathological' and a threat to the population. This lifestyle will not escape micro choices of self-formation, as stated by Bauman, in 'the land of the individual freedom of choice the option to escape individualization and to refuse participation in the individualizing game is emphatically *not* on the agenda' (2000: 34). Whether political or apolitical, individual or communal, or subversive or complicit, lifestyle serves to form and shape daily choices and activity of the subject. The practice of freedom of choice in the lifestyle *dispositif* combines technologies of power and technologies of the self in 'strategies that seek to govern us, and the ethics according to which we have come to govern ourselves' (Rose 1999: 9). The new ethical politics of agonistic consumption can transform consumptive practices into what Michel de Certeau describes as tactics of 'ingenious ways in which the weak make use of the strong, thus lend a political dimension to everyday practices' (1988: xvii). These themes, however, are addressed more fully in Chapter 6. I now turn to the network of lifestyle guidance that stylizes and secures choice, the body and health.

## Notes

1  Debord describes two types of spectacle, the concentrated and the diffuse. Debord writes: 'The spectacle exists in a *concentrated* or a *diffuse* form depending on the necessities of the particular stage of misery which it denies and supports. In both cases, the spectacle is nothing more than an image of happy unification surrounded by desolation and fear at the tranquil center of misery' (Debord 1983: 63). This suggests that spectacle cannot simply be reduced to 'commodity', 'image' or 'consumer society' but that it describes social and political arrangements of society through the mechanisms of the image to provide the appearance of unity.

2  A further, but later example, of the significance of everyday objects and consumerism is evident in the 'Yuppie' of the 1980s. According to Sam Binkley the rise of the transition from a 'loose lifestyle' of the 1960s counterculture to the entrepreneurial lifestyle of the 'Yuppie' in the early 1980s demonstrates the importance of everyday consumption. Increasingly individuals sought to express themselves through product choice and lifestyles such that the 1960s 'counterculture lost steam and the young

lifestylists were replaced by far more politically and morally innocuous yuppies' (Binkley 2007a: 99).

3  In addition to the broad use of the term ethics, Foucault says morality 'designates the manner in which [individuals] comply more or less fully with a standard of conduct… the manner in which they respect or disregard a set of values' (Foucault 1992: 25). This implies that ethics can be regarded as the conduct, while morality is the measurement of compliance of behaviour or practice to the rule of conduct. In the following chapter, I demonstrate the way the milieu of health information and imperatives to adopt a 'healthy lifestyle' serves to 'moralize' health and provide a measure against which the ēthos or lifestyle of the individual is compared.

4  Although Pierre Hadot has disputed parts of Foucault's analysis of Greek self-cultivation (Hadot 1995b), my interest is not the veracity of Foucault's account but his four axes of subject-formation as a guide in examining contemporary notions of lifestyle in the neoliberal governance of obesity.

# References

Abel, Thomas, and William C. Cockerham. 1993. "Lifestyle or Lebensführung? Critical remarks on the mistranslation of Weber's 'Class, Status, Party'." *The Sociological Quarterly* 34 (3):551–556.

Baehr, Peter. 2001. "The 'iron cage' and the 'shell as hard as steel': Parsons, Weber, and the Stahlhartes Gehäuse metaphor in the Protestant ethic and the spirit of capitalism." *History and Theory* 40 (2):153–169. doi: 10.1111/0018-2656.00160.

Bauman, Zygmunt. 2000. *Liquid modernity*. Malden, MA: Polity Press.

BBC News. 2008. *Chocolate "may cut diabetes risk"*. [cited July 16 2013]. Available from http://news.bbc.co.uk/2/hi/health/7363004.stm.

Binkley, Sam. 2007a. *Getting Loose: Lifestyle Consumption in the 1970s*. Durham, NC: Duke University Press.

Binkley, Sam. 2007b. "Governmentality and lifestyle studies." *Sociology Compass* 1 (1):111–126.

Bourdieu, Pierre. 1984. *Distinction: A Social Critique of the Judgement of Taste*. Cambridge, MA: Harvard University Press.

*Concise Oxford Dictionary of Sociology.* 1994. "Life-style." Edited by Gordon Marshall. Oxford: Oxford University Press.

Crary, Jonathan. 1989. "Spectacle, attention, counter-memory." *October* 50:97–107.

de Certeau, Michel. 1988. *The Practice of Everyday Life*. Translated by Steven Rendall. Berkeley, CA: University of California Press.

Debord, Guy. 1983. *Society of the Spectacle*. Detroit, IL: Black and Red.

Debord, Guy. 1998. *Comments on the Society of the Spectacle*. Translated by Malcolm Imrie. London: Verso.

Department of Health. 2011. *Change4Life – Eat Well, Move More, Live Longer*. [cited January 23 2012]. Available from http://www.dh.gov.uk/en/Publichealth/Change4Life/index.htm.

Elden, Stuart. 2004. *Understanding Henri Lefebvre*. London: Continuum.

Featherstone, Mike. 1987. "Lifestyle and consumer culture." *Theory, Culture and Society* 4:55–70.

Featherstone, Mike. 1991. *Consumer Culture and Postmodernism*. London: Sage.

Foucault, Michel. 1983. "On the genealogy of ethics: an overview of work in progress." In *Michel Foucault: Beyond Structuralism and Hermeneutics*, edited by Hubert Dreyfus and Paul Rabinow. Chicago, IL: University of Chicago Press.

Foucault, Michel. 1991. *Discipline and Punish: The Birth of the Prison*. Translated by Alan Sheridan. London: Penguin.

Foucault, Michel. 1992. *The Use of Pleasure: The History of Sexuality Volume 2*. Translated by Robert Hurley. Harmondsworth: Penguin Books.

Foucault, Michel. 2000a. "The ethics of the concern for the self as a practice of freedom." In *Ethics: Subjectivity and Truth*, edited by Paul Rabinow. London: Penguin.

Foucault, Michel. 2000b. "Technologies of the self." In *Ethics: Subjectivity and Truth*, edited by Paul Rabinow. London: Penguin.

Foucault, Michel. 2008. *Birth of Biopolitics: Lectures at the Collège de France, 1978–79*. Translated by Graham Burchell. Edited by Michel Senellart, François Ewald and Alessandro Fontana. Basingstoke: Palgrave Macmillan.

Giddens, Anthony. 1991. *Modernity and Self-Identity: Self and Society in the Late Modern Age*. Stanford, CA: Stanford University Press.

Hadot, Pierre. 1995a. *Philosophy as a Way of Life: Spiritual Exercises from Socrates to Foucault*. Edited by Aronold I. Davidson. Oxford: Blackwell.

Hadot, Pierre. 1995b. "Reflections on the idea of the 'cultivation of the self'." In *Philosophy as a Way of Life*, edited by Arnold I. Davidson. Oxford: Blackwell Publishing.

Heyes, Cressida J. 2007. *Self Transformations: Foucault, Ethics, and Normalized Bodies*. New York: Oxford University Press.

Howell, Jeremy, and Alan Ingham. 2001. "From social problem to personal issue: the language of lifestyle." *Cultural Studies* 15 (2):326–351.

Hoy, David Couzens 1999. "Critical resistance: Foucault and Bourdieu." In *Perspectives on Embodiment*, edited by Gail Weiss and Honi Fern Haber. New York: Routledge.

Jagose, Annamarie. 2003. "The invention of lifestyle." In *Interpreting Everyday Culture*, edited by Fran Martin. London: Arnold.

Klein, Naomi. 2002. *No Logo*. New York: Picador.

Larsen, Lars Thorup. 2011. "The birth of lifestyle politics: the biopolitical management of lifestyle disease in the United States and Denmark." In *Governmentality: Current Issues and Future Challenges*, edited by Ulrich Bröckling, Susanne Krasmann and Thomas Lemke. New York: Routledge.

Lasn, Kalle. 1999. *Culture Jam: The Uncooling of America*. New York: Eagle Brook.

Lazzarato, Maurizio. 2009. "Neoliberalism in action." *Theory, Culture & Society* 26 (6):109–133. doi: 10.1177/0263276409350283.

Lefebvre, Henri. 1996. *Critique of Everyday Life: Volume 1*. Translated by John Moore. New York: Verso.

Lemke, Thomas. 2011. *Biopolitics: An Advanced Introduction*. New York: New York University Press.

Let's Move. 2011. *Let's Move*. White House [cited January 23 2012]. Available from http://www.letsmove.gov/.

Mayes, Christopher. 2014. "Governing through choice: Food labels and the confluence of food industry and public health to create 'healthy consumers'." *Social Theory and Health* 12 (4):376–395.

Mayes, Christopher, and Donald B. Thompson. 2014. "Is nutritional advocacy morally indigestible? a critical analysis of the scientific and ethical implications of 'healthy' food choice discourse in liberal societies." *Public Health Ethics* 7 (2):158–169. doi: 10.1093/phe/phu013.

McNay, Lois. 1994. *Foucault: A Critical Introduction*. Oxford: Polity Press.

McNay, Lois. 2009. "Self as Enterprise: dilemmas of control and resistance in Foucault's *The Birth of Biopolitics*." *Theory, Culture and Society* 26 (6):55–77.

Merrifield, Andy. 2006. *Henri Lefebvre: A Critical Introduction*. New York: Routledge.

Moore, Robert. 2008. "Capital." In *Pierre Bourdieu: Key Concepts*, edited by Michael Grenfell. Stocksfield: Acumen.

O'Leary, Timothy. 2003. *Foucault and the Art of Ethics*. London: Continuum.

Plant, Sadie. 1995. *The Most Radical Gesture*. New York: Routledge.

Rose, Nikolas. 1999. *Powers of Freedom: Reframing Political Thought*. Cambridge: Cambridge University Press.

Sharpe, Matthew. 2003. "The logo as fetish: marxist themes in Naomi Klein's *No Logo*." *Cultural Logic: An Electronic Journal of Marxist Theory & Practice* 6. http://clogic. eserver.org/2003/sharpe.html

Shields, Rob. 1999. *Lefebvre, Love and Struggle: Spatial Dialectics*. New York: Routledge.

Thomson, Patricia. 2008. "Field." In *Pierre Bourdieu: Key Concepts*, edited by Michael Grenfell. Stocksfield: Acumen.

Weber, Max. 2001. *The Protestant Ethic and the Spirit of Capitalism*. Translated by Talcott Parsons. London: Routledge Classics.

Wedick, Nicole M., An Pan, Aedín Cassidy, Eric B. Rimm, Laura Sampson, Bernard Rosner, Walter Willett, Frank B. Hu, Qi Sun, and Rob M. van Dam. 2012. "Dietary flavonoid intakes and risk of type 2 diabetes in US men and women." *American Journal of Clinical Nutrition* 95 (4):925–933. doi: 10.3945/ajcn.111.028894.

Žižek, Slavoj. 2009. *Violence: Six Sideways Reflections*. London: Profile Books.

# 5 A cacophony of guidance

## Hearing, seeing and judging choices

> I believe that if we take care of our bodies and minds, we can realise our dreams…
> and never get old.
>
> Natalie Filatoff, editor of *Prevention: Smart Ways to Live* (2009b)

In May 2012, *Forbes* published an article entitled 'Why the U.S. May Go Broke over the Obesity Crisis' (Walton 2012). In many respects this is an unremarkable piece of commentary on a university study forecasting the economic impact of obesity in the US. A litany of government reports, research papers, television programmes and newspaper articles over the past decade have proclaimed the perils of obesity for population health and the economy. The remarkable feature of the article is its unremarkable advice – 'the tried and true lifestyle changes like diet and exercise, are probably the best bets for continuing to reverse the unnecessary epidemic' (Walton 2012). In Australia, the US and the UK, an obesity crisis purportedly threatens the very stability of society and calls for simple lifestyle changes as the 'best bets' to halt the crisis are virtually indisputable.

A common feature of these types of articles is their function as a non-traditional source of health guidance that emphasizes individual lifestyle changes to halt the obesity epidemic. Economists, journalists, personal trainers, ethicists, marketers and politicians are often the loudest in extolling the virtues of dieting and exercise. The explosion of lifestyle discourse in spaces away from the control of state-based health authorities mirrors Foucault's analysis of the transition from sovereign power to pastoral power and biopower. Surgeon generals, health ministers and physicians still have authority, but it is shared with and contested by many others.

The communication of lifestyle guidance is central to the network that enables governance of choices and bodies. Foucault's ideas on confession in the pastoral relation are useful for understanding the technologies of power and practices of the self that operate through lifestyle communication. During the eighteenth century the confession as a technique for producing truth led to its expansion from the ecclesiastical context into areas such as education, courts, the clinic and the home. The confession became a 'solemn rite' in the most ordinary and everyday parts of social life (Foucault 1998: 59). The expansion of the confession is evident in disciplines of law and psychoanalysis, but it has become increasingly popular in talk shows, reality television, lifestyle magazines and social networking websites like Facebook and Twitter. A variety of mediums have opened up avenues for individuals to confess their everyday lives and choices to peers, purported experts and authorities. These trends reinforce Foucault's observation that Western society has 'established the confession as one of the main rituals we rely on for the production of truth' (1998: 58). This is reflected in certain institutions, practices and relationships that employ the technique of the confession to elicit truth or desired information, ranging from the trivial to the profound.

In communication of lifestyle advice, the technologies of the self and technologies of power come into tension. Directives to monitor diet, quit smoking, reduce alcohol consumption and regularly exercise are held in connection with enticements of fat-free chocolate, diet pills, ab-toning workouts and swimsuit bodies. The commercialization of health blurs the lines between advertising and government campaigns to create a milieu of lifestyle information that provides new ways for individuals to read and recognize norms of the body, health and beauty. The rhetoric of lifestyle is imbued by a 'do-it-yourself consumerist health economy' that educates and shapes the public understanding of health and disease in ways beyond the scope of government agencies or medical professionals (Howell and Ingham 2001: 342). The physician's guidance is heard alongside the gym instructor's and the health food store clerk's.

Corresponding to the multiplication of government and commercial lifestyle guidance has been a growth in authorities, experts and guides. Government health campaigns and medical professionals undeniably continue to be a major source of health information that sets the tone of lifestyle discourse. However, more intimate and direct avenues of lifestyle communication between an individual and an expert have the potential to have sustained impact on the subject. In relationships with present or virtual experts, the subject incorporates 'official information in a creative and active way, within the contexts of everyday life, and against the backdrop of personal and shared experiences' (Bury 1994: 29). A nutritionist in *People* magazine may extol the benefits of a five-week carb-free diet to lose life-threatening fat and also have a body ready for summer, while on HBO's *The Weight of the Nation* documentary a different expert rebuffs carb-free diets and fast weight loss, suggesting instead a re-structuring of your eating habits and social relations to lose life-threatening fat (Hoffman et al. 2012).

The tension between different sources of advice does not undermine lifestyle guides, but reinforces the need for a caring pastor to navigate the terrain toward health and a firm body. The previous chapter introduced the idea of agonistic consumption to describe the way subjects are constituted through fragments of everyday practices directed towards aesthetic and biopolitical ends. This chapter continues to explore these tensions by addressing the cacophony of voices that issue lifestyle guidance in print and social media, television, and mobile devices. Building on previous chapters, I demonstrate the way medical and non-medical experts function as biopolitical pastors who hear confessions, circulate knowledge and deploy techniques to shape subjects in the lifestyle network. Biopolitical pastors occupy a space between Foucault's work on discipline and care, with individuals willingly placing their lives under the guidance of these experts to assist them in developing a healthy lifestyle or identifying bodies and practices as pathological. These pastoral relationships reinforce technologies of the self and the technologies of power that operate in the lifestyle network to govern and make visible everyday activities purported to be associated with obesity.

## Smartphones, social media, and technology

The Internet has become an indispensable part of modern life and has been instrumental in the rapid circulation of lifestyle knowledge, advice and guidance. Some government officials and health professionals have expressed dismay at the difficulty of regulating Internet-based advice or products related to health. Others view it as a useful tool. Dr Farooq Qureshi writes in a 'Practice tip' article for the *Australian Family Physician* that 'time constraints of a consultation meant that information must be provided quickly, accurately and in a form that is easily understood by the patient. Using a search engine such as Yahoo! makes this much easier' (Qureshi 2007: 538). In response, the Director of Preston Medical Library at the University of Tennessee, Sandy Oelschlegel, expressed dismay at Dr Qureshi's advice, stating, 'I do not understand how one could accept Yahoo! search results as the best choice...It is important to promote the use of best evidence in medicine. This article suggests using a tool that will not result in "best evidence" being identified' (Oelschlegel 2007: 982). The role of the Internet for patient and physician interactions is an important area of inquiry, and this exchange demonstrates suspicion that *even* physicians are incapable of navigating the myriad of health information on the Internet. The availability of knowledge and information via the Internet can challenge traditional sources of authority and alter patient and physician relationships. My interest is in the ubiquity of the Internet and its use as a source of health, nutrition and fitness information and as a means to share, generate and communicate health information.

The Internet is a field in which information and guidance is contested. On the one hand it is praised for challenging 'elitist' barriers around information put up by universities, newspapers or publishing houses (Poster 2001). On the other hand, the Internet is also considered a potentially dangerous source of health guidance that at best results in 'cyberchondria' or at worst serious misdiagnosis. Tania Lewis argues, 'unofficial health websites' have produced 'anxiety about the contested nature of medical authority on the web' (Lewis 2006: 527). Despite debates around the reliability of the Internet, it is a medium in which binaries of authorized and unauthorized or expert and non-expert sources of knowledge are blurred, with both contributing to the biopolitical milieu surrounding and informing choice, activity and lifestyle. The effect of the Internet means that the *Measure Up* or *Change4Life* websites containing government-endorsed lifestyle advice are in competition with the quick slimming diet pills promising overnight success or the more ambiguous industry-sponsored nutrition blogs and websites. The 'official' or 'authorized' advice of government or institutional bodies can be challenged or subverted through 'unofficial' discourses, or what some commentators refer to as 'cyber-quackery' (Iserson 2001). While the debates surrounding authority and legitimacy of the Internet are significant, I contend that these binaries do not account for the pastiche of lifestyle guidance and information surrounding the individual.

The explosion of health and fitness apps for smartphones and the increasing capabilities of wearable devices intensify individual engagements with Internet-based lifestyle information and guidance. In 2013 *mHealth Watch,* a website focused on mobile health or mHealth, estimated that there were '31,000 health, fitness, and medical related apps on the market' (Essany 2013). They also reported that 1 in 3 doctors recommend health apps to their patients (Essany 2015). These apps measure and quantify everything from caloric intake and exercise regimes to blood pressure and skin cancer risk. An article in *Wired* reports that the Food and Drug Administration (FDA) is facing a series of difficulties in gaining oversight, particularly of apps that purport to provide diagnostic tests through urine analysis via a smartphone camera (Copeland 2013). There is on-going debate among medical professionals and researchers as to whether these apps provide important tools for establishing healthy behaviours or another opportunity for individuals to be misled. Danielle E. Schoffman and co-workers believe apps 'could be powerful tools in the prevention and treatment of paediatric obesity' (Schoffman et al. 2013). Researchers also believe that such apps could help to address problems associated with self-reporting. Dhurandhar and colleagues argue that much of the data used to make dietary recommendations is flawed due to reliance on individual's reporting on their own eating and exercise practices (Dhurandhar et al. 2014). They suggest that more objective measuring devices are needed. These could include self-monitoring and self-tracking watches and other wearable devices. Emily R. Breton and co-workers, however, are concerned that 'many smartphone apps may lack comprehensive evidence-informed recommendations or practices for healthy weight control' (Breton, Fuemmeler, and Abroms 2011: 523). Companies such as Apple Inc. and Google are investing heavily to develop apps and devices. Working with biomedical device engineers, these companies are attempting to create products that will be able to provide health-related data for individuals, health care professionals and medical researchers (Gurman 2013). Whether or not these apps and devices will actually improve the health of the users, it is clear that they have a profound influence as a practice of subject formation.

The Quantified Self (QS) movement is a clear example of the way apps and smart devices are used to know, construct and modify the self (Lupton 2014, 2013). Gary Wolf is a journalist and co-founder of QS, which has the tagline 'self knowledge through numbers' (Quantified Self 2012). According to Wolf, self-tracking data can be used for a number of purposes, including public health and biosecurity. His interest, however, is self-knowledge. Wolf concludes his popular TED talk in saying 'the self is just our operation centre, our consciousness, our moral compass. So if we want to act more effectively in the world, we have to get to know ourselves better' (Wolf 2010). For the QS movement, knowledge of the self comes in the form of numbers garnered from self-tracking data. Using apps and wearable devices, people gather data about their sleep patterns, alcohol and caffeine consumption, caloric intake, energy expenditure, mood, heart rate, blood pressure and a range of other biometric information. These apps and devices enable people to quantify, calculate and measure daily choices and activities in ways not previously possible.

In addition to individuals voluntarily using these devices and apps, QS advocates believe government health agencies and services could operate more efficiently if the data gathering principles of QS were implemented. Roger Taylor, a former finance journalist, co-founded Dr Foster, a company that 'works with healthcare organisations to achieve sustainable improvements in their performance through better use of data' (Dr Foster 2014). According to Taylor, the use of data by individuals and health providers will reduce health expenditure and improve population health. However, the diseases and health issues that concern Taylor and other QS advocates reveal an extremely individualistic understanding of public health. In a BBC 4 radio programme on the quantified self, Taylor stated, 'Today, if we ask "what are the big threats to human health?" they're no longer microorganisms and external things. They are all to do with our own behaviour, and our own inability to control our behaviour' (Saunders 2013). Taylor and the QS movement believe that with more data individuals will be able to control their lives. It follows from this perspective that if you fall ill it is because you didn't have the data and/or the willpower to control your behaviour. Leaving aside the fact infectious disease remains a significant threat to human health, the QS perspective ignores the social and structural determinants of health. It also assumes widespread access to smart devices and the literacy required to use and understand the information they provide. This is not to suggest that there is no place for QS. However, considering the influence of advocates on government health departments, it is concerning that QS could further the process of reducing welfare to privatized and individualized choices (Pollock 2005, Taylor 2013).

In research for this chapter, I began using an app called The Eatery, which is described by the creators as allowing you to 'learn about yourself and your silent habits' (Massive Health 2013). I came across The Eatery in a review in the 'Health & Family' section of *Time*. According to Andrew Rosenthal chief strategy officer of the company that created the app, 'the goal of The Eatery…is to give people a leg up on eating better…[and] helping you change your fattening ways' (Park 2012). The app works by taking a picture of every piece of food you eat and then you rate the food using 'a healthiness meter' which is a spectrum from 'fat' to 'fit' (Park 2012). It is ambiguous whether 'fit' or 'fat' are describing the food itself or the projected future state of the consumer. The app then shares the picture of the food with other users who judge whether the meal is (or you are) 'fit' or 'fat'. The app aggregates the scores from each meal, along with the time and location in which it was eaten, and then produces a weekly analysis of how healthy you are eating. The weekly average is used to compare with past weeks and also with other users.

At the time of writing I have snapped 300 foods. I am currently eating '79% healthy'. In the week of June 10, I ate '73% healthy', which was '78% better than other people'. My 'healthiest' meal was at a Korean restaurant. My 'unhealthiest' meal was a piece of birthday cake at my parents. Automated daily messages of encouragement are also sent, such as 'Two eggs was rated 91%. 41 day streak! Digital high five!' Other users can comment or 'like' food. The app can also be synched with Facebook so that you can find friends using the app and have a more detailed profile in the Eatery app.

These techniques could be read as the most recent example in a long history of interrogating and routinizing everyday life. Smartphones have enabled a much more intense and public examination of food choices. The logic of the app is that others are better and more honest at judging the healthiness of a meal than the individual. Unlike a food diary, The Eatery is a public account that depends on unknown others assessing the healthiness of food choices. Innocuous habits such as snacking on crackers with peanut butter are not only judged against an individual's purported values of nutritional health but through The Eatery these practices are willingly offered for judgement. The article in *Time* notes that the 'key to the app is other users, who keep you honest' (Park 2012). Not unlike the penitent turning to the confessional, the value of using The Eatery is being judged and scrutinized by others.

The Eatery, and the thousands of other apps like it, cannot be completely theorized as a biopolitical pastor. Unlike the relation between a patient and physician or client and life coach, the relations enabled via these apps are plural and the others are unknown. However, there are features that overlap with the pastoral relation that hears a confession and guides the penitent. Foucault describes the confessional relation as requiring 'the presence (or virtual presence) of a partner who is not simply the interlocutor but the authority who requires the confession, prescribes and appreciates it, and intervenes in order to judge, punish, forgive, console, and reconcile' (Foucault 1998: 61). In the case of smartphone apps there is a collective of virtual partners waiting to hear (or see) the confession of food choices. The authority of this pastor is not based on scientific knowledge but the collective. The Eatery offers a collective partner that is eager to intervene, judge and help 'you change your fattening ways' (Park 2012). The statistical comparison to one's past eating habits and the eating habits of others establishes a relation between the individual and the collective that further reinforces ideas of success or failure in the practices of everyday life. The collective intervention of The Eatery encourages and enables the user to be vigilant over food choices and their relation to health and the body.

Although the newness and technical novelty of mHealth and smartphone apps make them alluring for users, marketers and researchers, it would be a mistake to ignore more traditional forms of lifestyle guidance. Magazines, television programmes and celebrities continue to play an important role in shaping the way health, bodies and conditions such as obesity are understood.

## Lifestyle media: 'smart ways to live well'

A major contributor to the network of lifestyle guidance and processes of subject formation is 'lifestyle media'. Lifestyle media has steadily occupied a larger section of television programming, newsstands and the Internet. A glance at any major newspaper or its corresponding website reveals a section devoted to 'lifestyle' that purports to be a serious source of the latest research on dieting and exercise geared towards weight-loss. For example, the *New York Times* website includes a BMI calculator in the 'healthy guide' section (*New York Times* 2011). Bell and Hollows suggest 'lifestyle media can be understood as guides to what and how to consume, and select from, a vast array not only of goods but also of services and experiences' (2006: 4). Unlike government health campaigns, lifestyle media presents a variety of information and draw on a plethora of 'experts' to lend authority to claims regarding health, nutrition, exercise, consumption, fashion and cosmetics.

In addition to providing a diversity of experts, lifestyle media enables the individual to 'read' and 'see' the lifestyle choices of others. The stated objective of much of the literature in the lifestyle genre is to guide (or pastor) individuals to improve their lives morally, aesthetically and physically (Bell and Hollows 2006: 4). Echoing Bourdieu, Bell and Hollows argue that the significance of lifestyle media is not simply in the promotion of particular lifestyles, but 'they also create the basis for making distinctive lifestyles legible' (2006: 5). Howell and Ingham suggest that the significance of lifestyle evaluations is that 'the [socially] evaluative component is also a morally evaluative component so that invidious social comparisons are also invidious moral comparisons' (2001: 346). Magazines, talk-shows, and DIY programmes provide a glossary through which lifestyle practices are legible, visible and communicative to others as signifying a particular form of life.

The legibility of lifestyle enables the moralization of health and associated choices. The moralization of health is implicit throughout lifestyle media and its focus on obesity. Although liberal societies are characterized by neutrality toward conceptions of the good life, health as defined by biomedical norms is increasingly considered a primary good (Crawford 1980, Lupton 1995). In biopolitical societies health becomes an undisputed good and the individual is responsible for maintaining this good (Mayes and Thompson 2014). This means that if food choice is perceived to diminish health – as determined by BMI, blood cholesterol level or other risk factors or biomarkers – it is by definition 'unhealthy' and falls outside of the 'good' of the individual and 'good' of the society. The guidance for individuals to choose 'good' health, 'beautiful' bodies and 'organic' food is emphasized in the rhetoric of responsibility and the idea these choices can be read by others as an indicator of a 'good life'.

It would be an impossible undertaking to catalogue or review the breadth of lifestyle media. This would also be unnecessary due to the highly repetitive nature of this literature. Consequently, I focus on the lifestyle magazine *Prevention: Smart ways to live well* to illustrate the biopolitical dynamics of this genre. *Prevention* is a US-based monthly magazine founded by Jerome Irving Rodale in 1950 in Pennsylvania, USA. Rodale was an entrepreneur, author and early advocate of organic gardening. *Rodale Inc* is now a global multi-platform content company that specializes in health and fitness and has a global reach to over 100 million people through distribution networks in 67 different countries. It is also partnered with *Reveille* and *NBC* to produce *The Biggest Loser* programmes and books (Rodale Inc 2015).

In 2009, *Prevention* was launched in Australia.[1] A survey of the first Australian edition of this magazine is illuminating as the editorial outlines the objectives and purpose of the magazine, demonstrating the Anglo-American appeal of lifestyle guidance to structure the everyday practices to prevent disease and promote health. While the purpose of *Prevention* is very different to official or governmental discourse found in campaigns like *Measure Up* and *Change4Life*, the tone is similar and there are a number of points of harmony between the guidance offered.

The editorial in the first edition states that *Prevention* aims to be a 'life-changing source of information to help you live your best life' (Filatoff 2009a: 11). *Prevention* is to assist individuals (specifically women over 40) to actively construct a lifestyle that promotes health. In drawing on 'news and advice on the latest scientific breakthroughs and the experience of leading health and medical experts' *Prevention* intends to help individuals develop lifestyles that resist the physical and psychological effects of ageing, reduce risk of all disease, and improve physical appearance (Filatoff 2009a: 11). Throughout *Prevention* the link among everyday activities, health status and the importance of lifestyle for promoting health, beauty and energy is emphasized. These claims are buttressed through an appeal to experts and recent studies. Examples include, 'women's-health specialist Emily Bradley [says] **"Diet and lifestyle play an enormous role in disease risk**, and we have control over these things"' (Filatoff 2009b: 12) (bold in original). And an article by Associate Professor David Colquhoun, 'Unchain Your Heart', states, 'research clearly shows that most heart attacks and other forms of coronary heart disease are due to poor lifestyle choices and habits. Do healthy-lifestyle factors really influence your risk of heart disease? You bet they do' (Filatoff 2009b: 35). As with government endorsed health promotion, the link between lifestyle choice and disease risk is reinforced in *Prevention* by rhetoric of 'expert knowledge' and the 'latest research'.

Weight-loss and dieting were a central theme of lifestyle media well before governmental concerns over the obesity epidemic. However, the mobilization of public health campaigns and widespread concern over the obesity epidemic, the lifestyle media's focus on weight-loss and a slender body, resonates with and is strengthened by the official public health rhetoric. The 'breakthrough' article on the cover of *Prevention* is titled: 'Flatten your belly: Lose 10cm in Just 4 weeks!' (Vaccariello and Sass 2009: 130). While the four-week time frame may attract concern from a physician or the architects of *Measure Up*, the goal to reduce centimetres from the waist accords with the objective of *Measure Up* and other health promotion campaigns. Echoing the objectives and rhetoric of anti-obesity discourse, the article argues that 'shrinking your waistline is not just about vanity: excess belly fat can significantly increase your risk of heart disease, the number one health threat to women, and increase your likelihood of developing other serious medical conditions' (Vaccariello and Sass 2009: 130). The cause of this situation is playfully described as 'your love of ice-cream, the irresistible aroma of hot chips, or the claustrophobia you get whenever you enter a gym!' (Vaccariello and Sass 2009: 132). The causal agents for increased waistline and body weight are everyday consumption and inactivity. Similarly the 'cure' is located in the everyday through the consumption of foods high in monounsaturated fatty acids – avocado, nuts, olives, dark chocolate.

Lifestyle media introduces a variety of 'experts' and guides to discussions of health, the body, nutrition and beauty. These experts complement the government discourse by providing different voices to the chorus of modifying choices to promote health. The *Prevention* article on weight-loss draws on the voice of the expert, but also the voice of the journalist and experienced peer. First, there is the light-hearted voice of the journalist using terms like 'tummy' or 'tum' and explaining scientific information with playful examples like 'the jiggly stuff just under the skin that makes it hard to zip up your jeans' (Vaccariello and Sass 2009: 132). There is also the more sophisticated scientific voice employing terms such as 'monounsaturated', 'subcutaneous fat' and 'metabolic syndrome', implying a certain gravitas to the claims about health risks and benefits of particular activities. And finally a third voice is the 'triumphant' peer who succeeds with the diet, employing positive and incentivizing phrases such as 'I'm feeling so much more energetic!', 'I'm shocked by how much I lost!' and 'My life needed a kick-start, and this was it!' (Vaccariello and Sass 2009 p. 142–145) The combination of these voices surrounds the reader with a plurality of guides within the one article that direct the individual toward the same end but using different language and tone.

This article is not exceptional. The following issue of *Prevention* has a very similar article on '5-minute Flat Belly Meals' (Gensler 2009). The interesting feature of these diets and others like them is the way they echo government health messages and employ a matrix of discourses to target and shape aspects of everyday life. Repeating the lesson of *Measure Up*, the 'scientific study' informs the reader that there is 'hidden fat' or 'dangerous stuff lurking inside' that leads to chronic disease (Vaccariello and Sass 2009: 132); but through the adoption of a particular regimen that monitors the type of food eaten, managing stress and emotional eating, the individual is able to secure and promote health. *Prevention* draws the reader's attention to the health significance of everyday practices such as tea or coffee drinking, hanging out the washing, and eating mushrooms or orange peel, with each reducing or increasing the risk of particular diseases. The last page of *Prevention*, entitled the 'Go! Guide' provides readers with 41 tips. At a quick glance the reader learns that red capsicum reduces wrinkles, grape juice reduces cholesterol, sunflower seeds lower heart disease risk by 23 per cent, or that the reader can 'drop a dress size by eating peanut butter, avocado and chocolate (really!)' (Filatoff 2009b). The emphasis on dropping a 'dress size', 'defying the conventions of ageing' and always 'feeling young and energetic' suggest a variety of norms which the individual is guided towards and encouraged to embody. The combination of health campaigns, lifestyle media and advertising establishes and circulates norms of femininity, beauty, body shape and the ageing process that allow readers to identify health promoting activities as well as those associated with disease and illness.

In her analysis of the female body in Western media and advertising, Susan Bordo demonstrates that (lifestyle) media contributes to a '*system* of values and practices within which girls and women – and increasingly men and boys as well – come to believe that they are nothing...unless they are trim, tight, lineless, bulgeless, and sagless' (2003: 32). The uncertainty and social anxiety of failing to conform to norms of femininity and masculinity, health and beauty mobilizes techniques of self-discipline and incentives (Monaghan 2008). The normalizing effect of lifestyle media generates notions of the 'healthy', 'beautiful', 'energetic' and 'intelligent' embodied subject. These pastoral techniques, knowledges and practices encourage individuals to place themselves in relation to the norm, such that the ability or inability to conform to the norm evidences the value of the subject. Cressida Heyes argues, 'the relationship we have to our embodied subjectivity is often assumed to speak to our value as ethical agents. Fatness declaims sloth, lack of discipline, greed, and failure to moderate appetite' (2007: 9). Through normalization the individual relates themselves to those around them and the wider population, and it becomes possible to 'see' and recognize the individual who conforms to, or deviates from, the position of the 'healthy woman' who 'takes care of her body and mind' and is 'excited about tomorrow' (Filatoff 2009a: 11).

*Prevention* is only an example of a wide variety of lifestyle media that targets and surrounds individuals, both male and female. It is important to emphasize that the individual is not a passive object that is seized and shaped by lifestyle media, but actively engages and seeks out information, advice and practices of life. Throughout *Prevention*, and lifestyle media in general, three themes are repeated: everyday practices are contested fields that can promote health or increase risk of disease; extensive research and expert knowledge confirms this link; and through changing everyday practice and developing a lifestyle an energetic, healthy and beautiful life is attainable. While government discourses can be disciplinary, lifestyle operates more with seduction and enticement. The tension between lifestyle media with government health campaigns constitutes the subject to govern choices and produce bodies that evidence a styled and secured life.

## The ordinary and exceptional guidance of celebrities

In 2010, Andrew Lansley, the UK Health Secretary, introduced public health initiatives designed to curb obesity and improve the health of the population. Known as the Public Health Responsibility Deal, Lansley sought to establish a partnership among the food and beverage industry, public health and government to design and implement public health policies related to food, exercise and alcohol (Secretary of State for Health 2010). The resulting policies and strategies have been heavily criticized by public health advocates for being weak measures that serve the interests of the very industries that many hold responsible for the obesity epidemic and diet-related chronic diseases (Lawrence 2010a, b, O'Dowd 2011). Surprisingly, one of the most vocal critics of the initiatives related to food and health has been celebrity chef Jamie Oliver. When Lansley suggested that people need to be 'honest about what they eat' and cut out 100 calories a day, Oliver was indignant. He described the suggestion as 'just worthless, regurgitated, patronising rubbish' (Ashton 2011). Oliver was not criticizing the strategy for unjustified government intrusion into people's lives or interfering in the restaurant and food industry of which he is a major beneficiary. Rather, Oliver felt that more substantial reforms were needed. 'Education is the key,' argues Oliver, 'We need to give people the knowledge to make better choices' (Ashton 2011). Echoing the discourse of public health and health promotion, Oliver considers education and empowerment as central to transforming the food choices and bodies of individuals and the population.

Like smartphone apps and lifestyle media, celebrities are increasingly contributing to the power/knowledge network of lifestyle as means to govern choices related to obesity. While apps and magazines provide knowledge and virtual pastoral relations, celebrity advocates like Oliver and Michelle Obama offer the possibility of a more intimate relation that directs choice and guides conduct. Oliver and Obama are leading voices in the chorus calling for public health reform and lifestyle change in the UK and the US.[2] Yet neither holds expertise relevant to advising government or individuals on nutrition and health.

Michel de Certeau's description of the expert helps to articulate how Oliver and Obama have become authorities on obesity, nutrition and lifestyle. Their authority does not rest on technical knowledge or official position, but is an extension of their speciality in other areas: chef, entrepreneur, self-made, First Lady, mother, lawyer. The exceptionalism in one area of life is translated or exchanged into authority in another area (de Certeau 1988: 7–8).[3] Through this process, Oliver and Obama, like *Prevention* and smartphone apps, have become pastors that care for the flock and guide the one and the many towards biopolitical ends of life and health.

Oliver's use of his celebrity and television programmes to change the eating habits of Britons, Americans and Australians to prevent obesity has resulted in his transformation 'from rock star celebrity to self-styled social activist' (Rousseau 2012: xi). Oliver's self-styled activism has produced a number of television programmes (*Jamie's School Dinners, Jamie's Ministry of Food, Jamie Oliver's Food Revolution, Jamie's Dream School,* and *Jamie's 15-Minute Meals*) most of which have book tie-ins. Oliver's diagnosis of the obesity epidemic is that people are becoming obese because they eat too much 'junk food', which they eat because they don't know how to choose or cook healthy food. His solution is to teach people how to identify, select and cook meals from scratch that promote health. *Jamie Oliver's Food Revolution* – (set in Huntington, West Virginia, US) – and *Jamie's Ministry of Food* – (in Rotherham, South Yorkshire, UK) – focused on Oliver revealing to school kids that chicken nuggets are 'fake' and 'disgusting', while simultaneously educating people how to cook nutritious meals from scratch. These programmes have been described as 'legendary' (Novick O'Keefe 2013) and according to a review in the *British Medical Journal* Oliver 'has done more for the public health of our children than a corduroy army of health promotion workers' (Rousseau 2012: 61).

Not everyone appreciates Oliver's approach.[4] However, with an estimated worth of $172m, Oliver's message clearly resonates with a significant section of consumers across the globe. Writing about Oliver's first transition from the 'mockney gobshite' of *The Naked Chef* to the 'saint' of *Jamie's Kitchen* (where he helps unemployed youths), Mark Lawson points to the tension within the persona of Oliver – 'It shouldn't be possible to be at the same time the face of Sainthood and the face of Sainsbury's but he has managed it' (2002). Similar criticisms were made following his expansion into Australia and partnership with the supermarket chain Woolworths (O'Chee 2014). Following his second transition from social worker to health promoter, Oliver now embodies the tension of food activist calling for revolution and food entrepreneur with a multi-million dollar suite of books, magazines, programmes and kitchenware. It is too simplistic to accuse Oliver of being a contradiction or a hypocrite. Oliver as revolutionary is made possible because of Oliver as entrepreneur, which is in turn authenticated and reinforced by Oliver as revolutionary. It is this dynamic that allows Oliver to be regarded as an expert guide in the daily activities and habits of the individual.

Similar to Oliver, Michelle Obama's authority as a lifestyle guide is ambiguous, yet she has enormous influence. Obama has made obesity – childhood obesity in particular – the defining issue of her role as First Lady. Obama has been involved in establishing legislation, including the Healthy, Hunger-Free Kids Act, which was signed into law by President Obama on December 13, 2010. However, Obama's primary role has not been legislative but normative. From the launch of her *Let's Move* initiative in 2010 to doing push-ups on *The Ellen DeGeneres Show* in 2012, her focus has been on the 'small changes that add up – like walking to school, replacing soda with water or skim milk, trimming those portion sizes a little – things like this can mean the difference between being healthy and fit or not' (Office of the First Lady 2010). It is the 'small changes' of daily life that are both within the control of the individual and produce the most significant results.

Obama does mention social factors beyond the control of the individual. She is concerned about the influence of industry, food deserts and the nutritional value of school lunch programmes. However, Obama does not want to address these issues with legislation or regulation, but in establishing norms for industry and government behaviour that increase the capacity for consumers to choose. For example, the food industry can assist by making 'food labels more customer-friendly', the beverage industry will 'be taking steps to provide clearly visible information about calories' and the entertainment industry will 'launch a nationwide public awareness campaign educating parents and children about how to fight childhood obesity' (Office of the First Lady 2010). Each of these industries has been accused of contributing to an environment that creates the conditions for obesity. However, Obama like Oliver considers the problem of childhood obesity to be a lack of education regarding the norms of health and nutrition. Too many parents and children are inactive, ignorant of basic nutrition and do not possess the necessary culinary skills to cook a meal at home. In Bourdieu's terms, it is the lifestyle choices not the social field structuring those choices that is the problem.

Obama's speech launching *Let's Move* is positively reviewed in the journal *Bariatric Nursing and Surgical Patient Care*. The review concludes, 'the solution to childhood obesity rests in the hands of parents and concerned adults, rather than legislation and executive orders' (Rowen 2010: 196). The *Let's Move* initiative and Obama's countless talk-show appearances and interviews in lifestyle media (including an interview in the US edition of *Prevention*) (Salvatore 2012) serve to circulate norms of health associated with a particular way of life – Mom's should take charge of their children's health, children should play outside and not watch TV, and families should eat dinner together (Office of the First Lady 2010). Obama's authority to direct food choices and health-related behaviour is undoubtedly entwined with her position as First Lady. However, in her *Let's Move* speech, Obama is keen to emphasize that she is from humble beginnings and that she knows what it feels like to be a working Mom. Obama places the extraordinariness of her current situation alongside the ordinariness of her previous life to reinforce her credentials as a guide for living. Like Oliver's culinary skills, Obama's relation to the President only serves as an exceptional platform from which she is able to transform into a friendly and caring guide on health, nutrition and parenting.

In their commentary on Foucault, Hubert Dreyfus and Paul Rabinow argue that the creation and establishment of various experts has led the individual to believe that 'one can, with the help of experts, tell the truth about oneself'(1983: 175). As discussed in Chapter 1, the pastor 'designates a very special form of power' (Foucault 1983: 214). This 'special form of power' has four characteristics: salvation, sacrifice, concern for the one and the many, and knowledge of the conscience (Foucault 1983). By entering into a pastoral relationship with the expert, by following their voices and guidance, the individual subjects himself or herself to the expert and is able to know the truth about the self. Submission to Oliver or Obama may include watching their television programmes, listening to interviews, buying cookbooks, planting a garden, cooking a meal, writing to a politician and informing a friend about the benefits of cooking food in the home. Although arguably more comprehensive and sustained than the virtual relations available via the Internet or magazines, there isn't a collective. And the celebrity guidance of Oliver and Obama still operates from a distance. Both are adept at establishing and styling an image of ordinariness and intimacy that connects with people via mass media (Rousseau 2012: 47–64). However, their pastoral guidance remains mediated. Screens, microphones, cameras, press agents, handlers and security guards separate them from the flock. They fulfil features of pastoral power, such as guiding through communication, establishing norms, dispositions of care and salvation in the guise of health. But the ability to know and direct the conscience is limited. It is here that a more traditional guide, the physician, is able to hear the confession and authoritatively direct health-related choices in everyday life.

## The physician as confessor

The need to know and direct the conscience of the individual requires the technique of the confession. Originating in the ecclesiastical contexts of the eighteenth century, the confession is distributed 'into the whole social body', imbuing 'the family, medicine, psychiatry, education, and employers' (Foucault 1983: 215). The confession gives the pastor access to the life of the individual and directs it toward salvation, or in the present situation, norms of health. Although commonly considered a technology of power, I contend that it is also a technology of the self – a mechanism of care and violence that is employed in a variety of contexts directed towards a disparity of ends. The proliferation of confession in Western societies has produced the Western individual as a confessing animal that desires and expects to confess.[5] In the virtual context of smartphone apps and more the traditional context of the clinic, the confession translates the practices of everyday life into discourse, enabling the dissemination, reinforcement and deployment of knowledge, truth and practice of the individual. In the patient and physician relationship, the confession gives the physician access to the patient's everyday life enabling the identification of pathological practices and guidance toward a healthy lifestyle.

Immersed in the network of health and lifestyle guidance, the individual does not approach the physician or other official sources of advice unaffected (Bury 1994, Davison, Smith, and Frankel 1991). Although much of the advice offered through lifestyle media does not require official medical oversight, when an individual does require further assistance, perhaps prompted by lifestyle media, then a physician is ordinarily the first point of contact. The network of lifestyle guidance furnishes the individual with knowledge about illness and health-related behaviours, as well as the language to describe their lifestyle prior to the medical encounter. Furthermore, the individual is likely to have expectations about the role of the health practitioner in providing lifestyle advice and guidance.

The family physician or general practitioner has historically been marginalized in public debate and policy surrounding population health (Lewis 2003: 17). However, the obesity epidemic has led to an emphasis on the family physician as a strategic figure in guiding individual lifestyles. The family physician becomes a central hub in a biopolitical network of surveillance, normalization, discipline and structuring of everyday life.[6] An example from Australia demonstrates this shift. In a press release on 27 September 2005, then Minister for Health and Ageing, Tony Abbott, stated that 'rather than write a prescription for pills, doctors will be encouraged to discuss ways to make healthier, long term lifestyle changes, and write prescriptions for regular exercise and eating healthier foods' (Abbott 2005). More recently, the House of Representatives Standing Committee on Health and Ageing's report on obesity in Australia, *Weighing it Up*, suggests that obesity requires an 'individual management plan', and physicians 'are an excellent resource in the treatment, management and prevention of obesity…as most Australians visit their GP each year and […] these visits would present an opportunity for the patient's height and weight to be measured and discussed' (Australia Parliament House of Representatives Standing Committee on Health and Ageing 2009: 62). Furthermore, the Committee recommends that physicians 'could play a significant role in collecting data on the prevalence of obesity in Australia and assist in the ongoing surveillance and monitoring recommended by the Committee' (2009: 63). This discourse characterizes the physician as a strategic agent in the response to obesity. Furthermore, the physician is considered not only as the 'first contact' but a 'credible source of preventive advice' (RACGP 2009: 4).

Professional associations, such as the American Medical Association, urge physicians 'to assess key dietary habits (e.g., consumption of sweetened beverages), physical activity habits, readiness to change lifestyle habits, and family history of obesity and obesity-related illnesses' (Rao 2008). Similarly, the Royal Australian College of General Practitioners, consider the physician's role to 'increase patient awareness of the issues and potential dangers of weight gain... [and] help educate patients about the need to change their lifestyle' (RACGP National Standing Committee 2006: 1). Complementing aspects of lifestyle media and the communication of Oliver and Obama, the physician tries to make the patient aware of the risks and dangers associated with 'weight gain' and the benefits of having a 'healthy weight', especially as individuals have 'unrealistic perceptions of their own weight' (2006: 2). Thus the physician not only assists the patient in using practices of the self to construct a healthy lifestyle, but in using techniques of power and discipline they try to convince the patient that they have a dangerous or pathological body caused by their lifestyle.

Physicians employ a range of strategies and techniques to shape their patient's behaviours. Writing in the *Australian Family Physician,* Egger, Pearson and Pal developed a questionnaire to help physicians guide individuals towards healthy lifestyles. The inquisitive feature of the questionnaire reflects the pastoral drive to know the conscience or in this case the daily practices of the individual or patient. Egger, Pearson and Pal argue that the use of a 'Diet, Activity and Behaviour Questionnaire' (DAB-Q) can assist physicians in 'detecting aspects of both diet and lifestyle likely to adversely affect the body weight of an individual' (2005: 591). Through this tool, the physician is able to 'customise a lifestyle prescription for the individual, potentially increasing the chances of successful weight loss' (Egger, Pearson, and Pal 2005: 592). Echoing the guidance issued in *Prevention,* the physician can use the questionnaire results to develop a regimen to encourage the patient to choose 'low energy dense fruit, beans or lentils' and 'decrease television viewing time' (Egger, Pearson, and Pal 2005: 593). Egger, Pearson and Pal conclude by arguing that the DAB-Q is a useful tool for diagnosing the most potent and easily changeable behaviours, satisfying 'a key component of behaviour change for weight loss prescription; limited and manageable initial change aimed at increasing motivation to proceed to more permanent lifestyle change' (Egger, Pearson, and Pal 2005: 593). Through the DAB-Q, and other confessional techniques I have discussed elsewhere (Mayes 2009), the relationship between patient and physician is positioned to guide the subject toward structuring everyday activity in order to promote health through prudential choice, and in so doing secure the population from the threat of the obesity epidemic.

Considering the time pressures and constraints faced by most physicians and primary health care professionals it is doubtful that many of these strategies can be implemented as the authors intend. Like the culinary guidance offered by

Oliver or the diets in *Prevention* the focus is not whether these strategies *actually* work in the sense of getting individuals to modify behaviours and reduce their waist circumference. Rather, the rhetoric and strategies produce norms and create an imperative for individuals to accept responsibility for their health and to self-govern. The inability of the individual to conform to these norms is not a failure on the part of the strategy, but on the individual's ability to responsibly self-govern. While the physician may not have the available time to extensively interview patients about lifestyle choices, the physician is characterized as guide or pastor who has the knowledge and authority to conduct the individual enmeshed in the lifestyle network toward taking responsibility for healthy choices and the adoption of a healthy lifestyle. However, it is important to be clear that the argument is not that the physician creates a wholly domineering or opportunistic relationship with the goal of controlling, dominating or oppressing the patient. The combination of the physician and lifestyle network use the patient's freedom of choice to arrange the disparate practices of everyday life into a lifestyle that promotes health and secures the population from economic costs of unhealthy choices. It is in the context of the objective of securing the population through the freedom of the individual that the physician is a *biopolitical* pastor.

The cacophony of lifestyle guidance and the diverse pastors and experts issuing advice contributing to the lifestyle network produces at least four outcomes requiring further consideration. First, it decentres medical knowledge and governance. Second, it makes the bodies and everyday choices of individuals visible for the individual and the population as determiners of health or disease. Third, it positions individuals as self-governors that are responsible for their own health and bodies as well as the collective health of the population. Fourth, it problematizes critique.

The decentring of knowledge is a central thread running through the lifestyle network. This is not to suggest that traditional authorities such as universities, professional associations or departments of health are regarded as irrelevant or the epistemological equals of *Prevention*, Yahoo! or The Eatery. Rather, the production and dissemination of health knowledge is not homogenous or under the control of the medical establishment. It is heterogeneous, contested and co-opted. This is particularly evident in the way contingent urgencies such as obesity coordinate the heterogeneity of these knowledges, giving them an apparent coherence that transmits an intelligible and audible imperative while allowing differences to remain.

The second outcome – the visibility of bodies and choices as indicators of health – allows medical professionals and the public to guide individuals towards norms and also exposes those that do not adhere to the norms. In transmitting and circulating the norms of health and the body associated with individual choices, the lifestyle network makes the bodies and everyday choices of individuals visible. The visibility of bodies and choices associated with individual and population health status enables individuals to compare their own bodies and lifestyles to others and the norm. This is particularly evident in the QS movement where the 'readability' of biometric data is regarded as producing objective knowledge of the self. Further, the visibility of bodies and choices as disease- or health-promoting makes it possible to mobilize strategies of governance. A key feature in these strategies of governance is to encourage individuals to accept responsibility and control for their health and bodies.

The third outcome – to produce individuals as self-governors responsible for their own health – is a key theme in neoliberal governmentality. In accepting responsibility, the individual absolves claims on the state or population to guarantee their health, and instead is called upon by the state and population to maintain their own health as an instrumental good for the population's security. The freedom of choice ties the individual to being responsible for that choice. The alternative to this situation, as characterized by neoliberal theory, is that if the government assumes responsibility for individual's health, then the individuals are required to hand over their freedoms to the government (Larsen 2011, Ayo 2012).

The final outcome of the cacophony of guidance in lifestyle network seeking to govern the obesity epidemic is that it problematizes critique. This is the focus of the final chapters, but a few remarks here serve to introduce the discussion. Critique is perhaps the most crucial of the four outcomes listed here as it has implications for the extent of the influence of the other three outcomes. In saying that the lifestyle network problematizes critique, I contend that multi-linear and entangled lines of lifestyle knowledge and relations of power defy direct critique, making it impossible to 'cut off the head of the king' and resist the norms of health and the body (Foucault 1998: 89). The norms do not emanate from a single locale but are established and circulated through a multiplicity of locales, including the individual themselves. The decentring of state-based medical knowledge or institutions, and proliferation of health and bodily norms through a plurality of avenues changes the mode of critique and resistance required. This is not to suggest that a Foucauldian analysis renders critique or resistance futile, but new grammars, practices and communities of critique are needed. These new modes of critique need to open the possibilities to ask 'how not to be governed *like that*, by that, in the name of those principles, with such and such objective in mind and by means of such procedures, not like that, not for that, not by them' (Foucault 1997: 44). In a gesture towards such a critique, I suggest that the task to resist 'being governed like that' and critique the 'objective in mind' cannot lie with the individual, but needs to take form in relation with others. Through relations with others, the possibility of new grammars of critique and norms of health and life is established. Importantly these norms are not the foundation of the community but emanate from it (Foucault 2000: 114–115). It is to these ideas that I now turn.

## Notes

1  The recent publication of *Prevention* in Australia allows me to analyse the first editorial and articles that claim to set the tone of later issues. In many ways *Prevention* is similar to the multiplicity of lifestyle media addressing health, beauty and nutrition. However, *Prevention* does have a strong focus on developing a whole lifestyle that *prevents* disease and promotes health.

2   Jamie Oliver and Michelle Obama's influence also extends to countries like Australia, New Zealand and South Africa (Rousseau 2012).
3   For a discussion of this passage in relation to Oliver see (Rousseau 2012: 62).
4   See a series of articles by Rob Lyons for *Spiked.*
5   In her genealogy of confession Chloë Taylor argues that the compulsion to confess has 'entered not only into the arts, most notably, and even into philosophy, but even more insidiously into politics, economics, the sciences, law and pedagogy, and finally into the desires and intuitions of the modern soul' (Taylor 2008: 67).
6   Between 2007 and 2011 the Department of Health and Ageing has committed $650.4 million to 'improve the quality and accessibility of primary health care services through the GP Super Clinic program' (Department of Health and Ageing 2011).

# References

Abbott, Tony. 2005. *Prescription for a Healthy Lifestyle.* [cited June 10 2011]. Available from http://www.health.gov.au/internet/ministers/publishing.nsf/Content/1B04491990 F841B5CA2570890009DFDA/$File/abb116.pdf.

Ashton, Emily. 2011. "Eat 100 less calories per day, Andrew Lansley tells Brits." *The Sun*, 13 October.

Australia Parliament House of Representatives Standing Committee on Health and Ageing. 2009. *Weighing it Up: Obesity in Australia.* edited by House of Representatives Standing Committee. Canberra: Printing and Publishing Office House of Representatives.

Ayo, Nike. 2012. "Understanding health promotion in a neoliberal climate and the making of health conscious citizens." *Critical Public Health* 22 (1):99–105. doi: 10.1080/09581596.2010.520692.

Bell, David, and Joanne Hollows. 2006. "Towards a history of lifestyle." In *Historicizing Lifestyle: Mediating Taste, Consumption and Identity from the 1900s to 1970s*, edited by David Bell and Joanne Hollows. Aldershot: Ashgate.

Bordo, Susan. 2003. *Unbearable Weight: Feminism, Western Culture, and the Body.* Berkeley, CA: University of California Press.

Breton, Emily R., Bernard F. Fuemmeler, and Lorien C. Abroms. 2011. "Weight loss—there is an app for that! But does it adhere to evidence-informed practices?" *Translational Behavioral Medicine* 1 (4):523–529. doi: 10.1007/s13142-011-0076-5.

Bury, Michael. 1994. "Health promotion and lay epidemiology: A sociological view." *Health Care Analysis* 2 (1):23–30. doi: 10.1007/bf02251332.

Copeland, Michael V. 2013. "F.D.A. can't hold back stream of mobile health apps." *Wired*, June 3.

Crawford, Robert. 1980. "Healthism and the medicalization of everyday life." *International Journal of Health Services: Planning, Administration, Evaluation* 10 (3):365–388.

Davison, Charlie, George Davey Smith, and Stephen Frankel. 1991. "Lay epidemiology and the prevention paradox: the implications of coronary candidacy for health education." *Sociology of Health & Illness* 13 (1):1–19. doi: 10.1111/j.1467-9566.1991.tb00085.x.

de Certeau, Michel. 1988. *The Practice of Everyday Life.* Translated by Steven Rendall. Berkeley, CA: University of California Press.

Department of Health and Ageing. 2011. *2010–11 G.P. Super Clinic Commitments.* [cited June 8 2011]. Available from http://www.health.gov.au/internet/main/publishing.nsf/ Content/pacd-gpsuperclinics-budgetcommitments.

Dhurandhar, N.V., D. Schoeller, A.W. Brown, S.B. Heymsfield, D. Thomas, T.I.A. Sørensen, J.R. Speakman, M. Jeansonne, and D.B. Allison. 2014. "Energy balance measurement: when something is not better than nothing." *International Journal of Obesity* 39(7), 1109-1113. doi: doi:10.1038/ijo.2014.199

Dr Foster. 2014. *About Us.* [cited 11 June 2015]. Available from http://www.drfoster.com/about-us/.

Dreyfus, Hubert, and Paul Rabinow. 1983. *Michel Foucault: Beyond Structuralism and Hermeneutics.* 2nd edn. Chicago, IL: University of Chicago Press.

Egger, Garry, Suzanne Pearson, and Sebely Pal. 2005. "Individualising weight loss prescription: A management tool for clinicians." *Australian Family Physician* 35 (8):591–594.

Essany, Michael. 2013. "Mobile health care apps growing fast in number." *mHealth Watch.* http://mhealthwatch.com/mobile-health-care-apps-growing-fast-in-number-20052/

Essany, Michael. 2015. "1 out of 3 Doctors Suggesting Mobile Health Apps to their patients." *mHealthWatch.* http://mhealthwatch.com/1-out-of-3-doctors-suggesting-mobile-health-apps-to-their-patients-25378/

Filatoff, Natalie. 2009a. "Excited about tomorrow". In *Prevention: Smart ways to live.* Sydney: Pacific Magazines.

Filatoff, Natalie, ed. 2009b. *Prevention: Smart ways to live.* Sydney: Pacific Magazines.

Foucault, Michel. 1983. "Afterword: the subject and power." In *Michel Foucault: Beyond Structuralism and Hermeneutics*, edited by Hubert Dreyfus and Paul Rabinow. Chicago, IL: University of Chicago Press.

Foucault, Michel. 1997. "What is critique?" In *The Politics of Truth*, edited by Sylvère Lotringer and Lysa Hochroth. New York: Semiotext(e).

Foucault, Michel. 1998. *The Will to Knowledge: The History of Sexuality Volume 1.* Translated by Robert Hurley. England: Penguin Books.

Foucault, Michel. 2000. "Polemics, politics, and problematizations." In *Ethics: Subjectivity and Truth*, edited by Paul Rabinow. London: Penguin.

Gensler, Tracy. 2009. "5-minute flat belly meals: fat-fighting meals on a plate". *Prevention*, 61 (6):102.

Gurman, Mark. 2013. "iWatch's novelty emerges as Apple taps sensor and fitness experts". *9to5Mac: Apple Intelligence.* http://9to5mac.com/2013/07/18/apple-stacks-iwatch-team-with-sensor-fitness-experts/

Heyes, Cressida J. 2007. *Self Transformations: Foucault, Ethics, and Normalized Bodies.* New York: Oxford University Press.

Hoffman, John, Judith A. Salerno, Alexandra Moss, Kelly D. Brownell, and Harvey V. Fineberg. 2012. *The Weight of the Nation: Surprising Lessons About Diets, Food, and Fat from the Extraordinary Series from HBO Documentary Films.* New York: St. Martin's Press.

Howell, Jeremy, and Alan Ingham. 2001. "From social problem to personal issue: the language of lifestyle." *Cultural Studies* 15 (2):326–351.

Iserson, Kenneth V. 2001. "Commentary: the (partially) educated patient: a new paradigm?" *Cambridge Quarterly of Healthcare Ethics* 10 (02):154–156.

Larsen, Lars Thorup. 2011. "The birth of lifestyle politics: the biopolitical management of lifestyle disease in the United States and Denmark." In *Governmentality: Current Issues and Future Challenges*, edited by Ulrich Bröckling, Susanne Krasmann and Thomas Lemke. New York: Routledge.

Lawrence, Felicity. 2010a. "McDonald's and PepsiCo to help write UK health policy." *The Guardian*, 12 November.

Lawrence, Felicity. 2010b. "Who is the government's health deal with big business really good for?" *The Guardian*, 12 November.

Lawson, Mark. 2002. "The fall and rise of Jamie." *The Guardian*, December 5.

Lewis, Milton J. 2003. *The People's Health: Public Health in Australia, 1950 to the Present*. Westport, CT: Greenwood Press.

Lewis, Tania. 2006. "Seeking health information on the internet: lifestyle choice or bad attack of cyberchondria?" *Media, Culture & Society* 28 (4):521–539. doi: 10.1177/0163443706065027.

Lupton, Deborah. 1995. *The Imperative of Health: Public Health and the Regulated Body*. London: Sage Publications.

Lupton, Deborah. 2013. "Quantifying the body: monitoring and measuring health in the age of mHealth technologies." *Critical Public Health* 23 (4):393–403.

Lupton, Deborah. 2014. "Self-tracking modes: Reflexive self-monitoring and data practices." Paper presented at the Imminent Citizenships: Personhood and Identity Politics in the Informatic Age Workshop, Australian National University. August 27. Accessed June12, 2015  https://www.academia.edu/8043441/Self-tracking_Modes_ Reflexive_Self-Monitoring_and_Data_Practices

Massive Health. 2013. *The Eatery*. [cited 18 April 2013]. Available from www.eatery. massivehealth.com.

Mayes, Christopher. 2009. "Pastoral power and the confessing subject in patient-centred communication." *Journal Bioethical Inquiry* 6 (4):483–493.

Mayes, Christopher, and Donald B. Thompson. 2014. "Is nutritional advocacy morally indigestible? a critical analysis of the scientific and ethical implications of 'healthy' food choice discourse in liberal societies." *Public Health Ethics* 7 (2):158–169. doi: 10.1093/phe/phu013.

Monaghan, Lee F. 2008. *Men and the War on Obesity: A Sociological Study*. New York: Routledge.

*New York Times*. 2011. "Health and Fitness Tools." Arthur Ochs Sulzberger, Jr. 2011 [cited May 12 2011].

Novick O'Keefe, Linda. 2013. "Using the kitchen as a catalyst for change." *Huffington Post*, June 3.

O'Chee, Bill. 2014. "Jamie Oliver burnt by Woolworths partnership." *Brisbane Times*, June18.

O'Dowd, Adrian. 2011. "Government's public health responsibility deal is met with scepticism." *BMJ* 342. doi: 10.1136/bmj.d1702.

Oelschlegel, Sandy. 2007. "Letters to the editor." *Australian Family Physician* 36 (12):982.

Office of the First Lady. 2010. *Remarks of First Lady Michelle Obama – Let's Move Launch*. The White House 2010 [cited January 18 2012]. Available from http://www. whitehouse.gov/the-press-office/remarks-first-lady-michelle-obama.

Park, Alice. 2012. "Who eats what? an app tracks diets around the world." *Time*, May 30.

Pollock, Allyson. 2005. *NHS plc: The Privatisation of Our Health Service*. London: Verso.

Poster, Mark. 2001. "The internet and the public sphere." In *Reading Digital Culture*, edited by David Trend. Oxford: Blackwell Publishers.

Quantified Self. 2012. *About*. [cited 19 May 2015]. Available from http://quantifiedself. com/about/.

Qureshi, Farooq. 2007. "How I use the internet." *Australian Family Physician* 36 (7):538.

RACGP. 2009. *Guidelines for Preventive Activities in General Practice*. 7th edn. Melbourne: Royal Australian College of General Practitioners.

RACGP National Standing Committee. 2006. *Overweight and Obesity*. South Melbourne: Royal Australian College of General Practitioners.

Rao, Goutham. 2008. "Childhood obesity: highlights of AMA Expert Committee recommendations." *American Family Physician* 78 (1):56–63.

Rodale Inc. 2015. *About Us*. [cited 11 June 2015]. Available from http://www.rodaleinc.com/content/about-us.

Rousseau, Signe. 2012. *Food Media: Celebrity Chefs and the Politics of Everyday Interference*. Oxford: Berg Publishers.

Rowen, Lisa. 2010. "Childhood obesity: words from the First Lady." *Bariatric Nursing and Surgical Patient Care* 5 (3):195–196. doi: 10.1089/bar.2010.9913.

Salvatore, Diane. 2012. "Michelle Obama talks let's move! and her healthy lifestyle". *Prevention*, September.

Saunders, Frances Stonor. 2013. "The quantified self: can life be measured?" *Analysis*: BBC Radio 4.

Schoffman, Danielle E., Gabrielle Turner-McGrievy, Sonya J. Jones, and Sara Wilcox. 2013. "Mobile apps for pediatric obesity prevention and treatment, healthy eating, and physical activity promotion: just fun and games?" *Translational Behavioral Medicine*:1–6. doi: 10.1007/s13142-013-0206-3.

Secretary of State for Health. 2010. *Secretary of State for Health's speech to the UK Faculty of Public Health Conference – 'A new approach to public health'*. [cited January 25 2012]. http://www.dh.gov.uk/en/MediaCentre/Speeches/DH_117280.

Taylor, Chloë. 2008. *The Culture of Confession from Augustine to Foucault: A Genealogy of the "Confessing Animal"*. London: Routledge.

Taylor, Roger. 2013. *God Bless the NHS*. London: Faber & Faber.

Vaccariello, Liz, and Cynthia Sass. 2009. "Lose 5 kilos in 4 weeks! flat belly diet!" *Prevention: Smart Ways to Live Well*. 1: 130–145.

Walton, Alice G. 2012. "Why the U.S. may go broke over the obesity crisis." *Forbes*, May 11.

Wolf, Gary. 2010. "The quantified self". In *TED*. Cannes: TED Conferences, LLC.

# 6 Styles of resistance

## The body, counter-conduct and critique

> Yet God knows that there are ideological traffic police around, and we can hear their whistles blast: go left, go right, here, later, get moving, not now…The insistence on identity and the injunction to make a break both feel like impositions, and in the same way.
>
> Foucault, 'For an Ethics of Discomfort' (2000b: 444)

Bodies, choices and lives are relentlessly viewed, critiqued and judged in relation to norms of health and beauty. Restrictive norms of beauty, fashion and associated norms of gender have been the object of critic and resistance for decades. Health, however, has attracted less criticism or creative resistance. Why resist health or a healthy lifestyle? The seemingly unquestionable goodness of adopting a healthy lifestyle and apparently indisputable badness of obese bodies partly demonstrates the effectiveness of the lifestyle *dispositif*. Like mothers, apple pie and freedom, it is uncontroversial to declare 'health is good!' Yet, health and the various declarations of its goodness often mask social norms that limit, exclude and devalue some lives and bodies. This chapter examines the possibility of resisting and critiquing health and the idea of a healthy lifestyle.

The pursuit of health, as demonstrated by a slender body, controlled eating and regular exercise, is a powerful goal that is widely valued. Any suggestion of opting-out, choosing alternative goals, or moving the markers of achievement is characterized as dangerous or irrational. In 2013, the *Journal of the American Medical Association* published an epidemiological study that found people who are obese grade 1 (BMI of 30≤35) had no increased risk of dying prematurely and overweight people may actually have greater life expectancy (Flegal et al. 2013). Walter Willett of Harvard School of Public Health was indignant. He described the research on NPR as 'really a pile of rubbish' and that 'no one should waste their time reading it' (Carter and Walls 2013). A UK National Obesity Forum representative told the BBC, 'It's a horrific message to put out at this particular time. We shouldn't take it for granted that we can cancel the gym, that we can eat ourselves to death with black forest gateaux' (Carter and Walls 2013). These responses to Flegal et al.'s research highlight the strength with which the norms of health in relation to bodies are held on to and that resistance to them can provoke vicious attacks. As such, resistance needs to be multifaceted.

This chapter explores the possibility of resistance within the lifestyle network. The lifestyle network enables the shaping and styling of everyday practices into subjectivities that adhere to biological and social norms. The successful adoption of these norms makes the individual visible as a healthy subject. However, the norms and strategies deployed throughout this network are not neutral or beneficial for all individuals. Failure to adhere to these norms can result in disciplinary mechanisms and withdrawal of protection and care. The confluence of neoliberalism, epidemiology and health promotion enables techniques such as the tape measure, BMI and smartphones to inscribe the individual-body with medical and social norms of health. The blurring of institutional (physicians and government programmes) and non-institutional (lifestyle media, health apps, and celebrities) guides opens up avenues of power/knowledge that constitute the subject to make healthy choices and shape their body. This network, however, is not a stable or closed system but a 'crisscrossed' and 'multilinear' tangle of lines 'that discloses points of existing and possible resistance' (Bussolini 2010: 92). This chapter turns to examine the possibility for individuals to critique and resist subjectification along the lines of the lifestyle network. Operating in and with the agonism between technologies of power and technologies of the self, I question whether it is possible to develop a lifestyle of resistance – a lifestyle that challenges the conceptions of the body and health circulated in anti-obesity discourse.

To address the possibility of resistance to strategies of lifestyle and anti-obesity discourse, I draw on 'Health at Every Size' (HAES) and Fat Studies research. HAES is an approach to health that does not focus on body-weight or dieting and focuses on being attuned to 'natural hunger' and doing pleasurable physical activity. It is a diverse approach drawing on a variety of disciplines and fields. I focus on the work of Linda Bacon (2010).[1] I demonstrate that resistance to norms of health and the body from a HAES perspective operates in a manner similar to ideas outlined by Foucault. In particular, I examine his analyses of bodily resistance to the *dispositif* of sexuality, counter-conduct to pastoral power and critique of governmentality. In proposing the possibility of critical and resistant responses within the lifestyle network I do not argue that we can get beyond lifestyle – we have no choice but to choose and those choices are signals to others of a mode or style of life. However, subversive and counter-conducting choices can construct a lifestyle that disrupts and disables the enabling network of lifestyle.

Corresponding to the multiple lines of power/knowledge that traverse everyday life and 'make visible' healthy subjects, resistance contests the everyday through 'restless and endless' activity and mobilizes communal relations of care. The communal and relational aspect of resistance corresponds and counters the social and political importance of conforming to norms of health and the body. Considering the biopolitical and social significance of norms, any attempt to resist lifestyle norms requires tactful and strategic negotiation of disciplinary and exclusionary effects of biopolitical governance. In avoiding or failing to conform to norms, the individual becomes increasingly visible as a target for more intense disciplinary biopolitical interventions, even to the point of exclusion and exposure to violence. As I argue in the next chapter, communal relations are therefore important for establishing avenues and lines of resistance to the lifestyle network that guides and shapes bodies and choices. First, however, I discuss the different modes of resistance available in Foucault's work.

## Resist, counter, critique

In the 1982–83 lecture series *The Government of the Self and Others*, Foucault overviews his work as a series of theoretical displacements. One of the displacements is freeing thought from a general 'Theory of Power'. Instead of thinking in capitalized terms of Domination, Sovereign and Power, Foucault attempts 'to bring out the history and analysis of procedures and technologies of governmentality' (2010: 42). This genealogical focus on governmentality, rather than a Theory of Power indicates both the operation of power and form resistance takes in Foucault's thought. Instead of opposing power per se, strategic resistance requires the negotiation of the historical and governmental imposition of norms. My focus is to develop a critique of the implicit and explicit demands and enticements that individuals need to make healthy choices. These demands and enticements occur through the lifestyle network of public health knowledges, neoliberal policies and consumer practices. In order to provide a critical response to this network, it is useful to outline three responses sketched by Foucault in relation to the *dispositif* of sexuality, pastoral power and governmentality: resistance of bodies and pleasure, pastoral counter-conduct, and critique as ēthos. A discussion of these three locales of resistance serves in articulating the resistant responses to the enabling network of lifestyle outlined in the second part of this chapter.

### *Irregular bodies*

The body is a crucial site of power and resistance. Much of Foucault's analysis in *The Will to Knowledge* is on the historical processes that shape bodies and produce particular subjects around ideas of sexuality. Foucault was not only interested in the production of subjects via relations of power, but also 'why' and 'how' these relations could be resisted. Foucault initially offers 'bodies and pleasures' as an opening through which human life may resist capture by biopolitical strategies of modernity. Taking aim at Freudian psychoanalysis and strategies seeking to control sexuality, Foucault states 'the rallying point for the counterattack against the deployment of sexuality ought not to be sex-desire, but bodies and pleasures' (1998: 157). Life exceeds attempts to normalize it, particularly through Freudian sublimation, and this excess is expressed in bodies and pleasures. Locating resistance in bodies and pleasures has generated a wide literature debating the merits and meaning of this statement. Drawing on these debates, I contend that the body itself has the potential to alter strategic games of power. However, bodies alone are not the sole drivers of resistance.

Foucault's genealogical approach reveals the attachment and inscription of history and politics on the body. In 'Nietzsche, Genealogy, History' Foucault addresses the relationship between history and the body, and the role of genealogy in articulating the relationship. Genealogy traces the mutual influence of body and history, exposing the body as 'totally imprinted by history and the process of history's destruction of the body' (Foucault 2000c: 375–376). This relation between the life of the body and history – or bio-history – is an oscillating relation of confluence, with each interfering with the other (Foucault 1998: 143). This characterization of the body has not been without its detractors.

Re-visiting the debates over Foucault's idea of bodies and resistance serves to identify the kind of resistance needed in response to the lifestyle network. Bodies and pleasures became a focus of early debates between Jürgen Habermas and Foucault scholars (McWhorter 1999). Habermas did not consider bodies and pleasures to be an adequate normative basis for resisting relations of power. For bodies and pleasures do not substitute for a normative framework and nor do they escape the all-pervasive reach of biopower. Following Nancy Fraser (1981, 1983), Habermas argues that Foucault's account of resistance draws 'its motivation, if not its justification, only from the signals of body language, from that nonverbalized language of the body on which pain has been inflicted, which refuses to be sublate into discourse' (1987: 285–286). Habermas also notes that Foucault's emphasis on resistance emanating from bodies due to biopower's possession of the body rather than the mind, takes on a 'vitalist' form 'without needing any normative foundation to do so' (1987: 283). Habermas asks 'why' on Foucault's account 'should we muster any resistance at all against this all-pervasive power circulating in the bloodstream of the body of modern society, instead of just adapting ourselves to it?' (Habermas 1987: 283–284). Habermas contends that if resistance is based in the body, in a vitalist philosophy, then the question of normative grounds for resistance is not solved but merely evaded. The lack of a normative framework in Foucault's genealogical method leads to blind resistance, where the subject is not guided or motivated to take a particular path in the conflict with mechanisms and strategies of power.

More recently, Johanna Oksala suggests that subjects become stripped of agency and determined through historical forces of power/knowledge. Subjects become 'ontologically tied' to the historical influence of the 'power/knowledge network' (2005: 106), which leads to the problematization of freedom, subjectivity and resistance. If the body is inscribed by historical contingencies and produced through power/knowledge, then the very possibility for the subject to freely act, resist or self-create is determined by the effect of history and the power/knowledge network.[2] Thus the individual is trapped in a vicious circle in which the resisting subject is merely the product of power/knowledge.

In response to this line of criticism, Foucault maintained that the forces of the body effect and alter the strategic network of power/knowledge. In the interview, 'Body/power', Foucault was asked 'Is the liberation possible without the *quadrillage* [network of forms of control]?' His response is worth quoting in full.

> Mastery and awareness of one's own body can be acquired only through the effect of an investment of power in the body: gymnastics, exercises, muscle-

building, nudism, glorification of the body beautiful. All of this belongs to the pathway leading to the desire of one's own body, by way of the insistent, persistent, meticulous work of power on the bodies of children or soldiers, the healthy bodies.

(Foucault 1980: 56)

An initial feature of Foucault's answer, that relations of power inscribe and invest the healthy body, echoes the discussion in Chapter 5 regarding the way individuals actively style, shape and design their bodies through diet, exercise and other regimens. In the anti-obesity context, the body is not only invested with power through techniques of the tape measure, diets or physician interviews, but also through aesthetic practices, which can resemble disciplinary techniques. Oksala describes this process of marking and inscribing as being both voluntary and involuntary (2005: 120). Thus the body is involuntarily marked through the normalizing effect of strategies, such as public health campaigns, BMI and physician recommendations. Yet, there is also a voluntary response through the aesthetics of lifestyle media, gym or smartphone technology.

Neither the individual nor the body is passive in the lifestyle network of power/knowledge. Both produce counter-effects that re-constitute and modify the network. As outlined in the opening chapter, life provokes a response from politics. In the case of obesity governance, the excessive body has become an 'urgent need' that requires governance and control. Biopolitical strategies react to the forces of the body and in turn affect the body. Foucault argues that once power has produced an effect in the body, 'there inevitably emerge the responding claims and affirmations, those of one's own body against power, of health against the economic system, of pleasure against the moral norms of sexuality, marriage, decency' (1980: 56). The effect of power on the body produces a counter-effect that can exceed the initial influence. However, the counter-effect does not permanently eradicate the influence of power. Strategic relations of power re-organize and move its focus elsewhere to continue the struggle (Foucault 1980: 56).

The counter-effect follows an agonistic or strategic logic such that 'after investing itself in the body' relations of power/knowledge 'finds itself exposed to a counter-attack in that same body' (Foucault 1980: 56). The power effect on bodies does not simply produce a docility that suffocates and precludes alternatives but incites the body to respond by transgressing norms. The response of 'bodies and pleasure' to the *dispositif* of sexuality and its moral norms takes the form of cultivating of pleasures as a bodily dividend responding to the initial investment of power. Medical and political interventions into sexuality not only order and define, but unwittingly produced the counter-effect of pleasurable 'perversions'. According to Bordo, the medical, scientific and juridical interventions 'ferreted out, eroticized and solidified all sorts of sexual types and perversions, which people then experienced (although they had not done so originally) as defining their bodily possibilities and pleasures' (2003: 142–143). Through psychiatric power and techniques of the confession, sexual perversions were targeted, seized and manipulated, but in the very process, new bodily possibilities and pleasures were produced.

The strategic logic also operates in the lifestyle *dispositif*. The body has become the object of intense scrutiny in the obesity epidemic. This is evident in the arguments of Dan Callahan, who believes the threat posed by obesity is so great that it justifies coercive and stigmatizing strategies that force obese individuals to 'want a good diet and exercise for themselves and for their neighbour and…[know] that excessive weight and outright obesity are not socially acceptable' (2013: 37). Callahan's view reflects a wider perspective that bodies outside the medically determined norms are justified targets of social and political coercion (Puhl and Heuer 2010, Puhl, Peterson, and Luedicke 2013, Goldberg and Puhl 2013).

A counter to the anti-obesity discourse is the HAES movement. Linda Bacon, a prominent figure in HAES, argues that the body is inscribed by 'every authority and institution [that] urges us to fight fat' (2010: xxiii). According to Bacon, anti-obesity discourses transform the body, weight and choices perceived as health-related from issues of public health into moral and social issues requiring action, guidance and intervention. 'From every pulpit of weight control' writes Bacon, 'we hear a singular message: Follow my plan, you'll lose weight, improve your health, become a better person, and have a happier life' (2010: xxiii).

The HAES movement rejects the pursuit of weight loss for the purpose of improving health. It proposes a 'weight-neutral approach [focusing] on loving self-care and the decisions that people can make on a day-to-day basis that are sustainable for a lifetime' (Burgard 2009: 44). The body and pleasure are central themes in HAES, with a focus on 'appreciating the wonderful diversity of body shapes', discovering the 'pleasure of eating well', developing a 'joy of movement' and acknowledging 'beauty and worth in EVERY body' (Burgard 2009: 42–43). The diversity, pleasure and the uncontrollability of the body form the ground through which HAES attempts to establish resistant bodies and new pleasures within the network of lifestyle governance (Burgard 2009: 51).

As a strategy of resistance, HAES attributes an agency to the body. According to Bacon, if individuals ignore dieting messages and listen to their bodies, their weight will stabilize at a natural 'setpoint'[3] representing the individual's ideal weight – 'your body will be guiding you in making nutritious, pleasurable choices' (2010: 13). The agency of the body is contrasted with external rationalizations that focus on 'counting calories, totalling fat grams, or weighing broiled skinless chicken breasts!' (Bacon 2010: 29). Adopting a kind of vitalism, HAES recommends abdicating control to the body, as 'biology is much more powerful than willpower' (Bacon 2010: xxv). The body instigates resistance against the norms produced through lifestyle media and the diet industry, but also serves to guide the 'poor chooser' toward nutritious and pleasurable food choices.[4] Further, Bacon and others in the HAES and Fat Studies literature argue that the obese body is the resistant product of the diet industry. Attempts to control the body through diet regimens are argued to have contributed to the production of the unruly and uncontrollable obese body with its associated conditions like cardio-vascular disease (Gaesser 2002, 2009, Bacon 2010: 49, O'Hara and Gregg 2006: 261).

The role of 'bodies and pleasure' in transforming the power/knowledge network is an important initial opening for resistance to occur. A further feature of the body as resistance is the genealogical demonstration of the instability of the body and the way it was lived. Genealogical analyses highlight the historical unrest and instability of the body.[5] The seizure of the body by power cannot be total or absolute, but is vulnerable to the flux of the body. The use of genealogy to demonstrate the historical inscription and malleability of the body points towards the power/knowledge effect on the body, but also to the unpredictability of the body and critique. The irregular and malleable character of the body serves as an avenue for both the operation of lifestyle norms and resistance to those norms.

According to David Hoy, Foucault's work on genealogy further develops the possibility of bodily critique. Hoy argues:

> If the body can be shown to have been lived differently historically (through genealogy), or to be lived differently culturally (through ethnography), then the body can be seen to be 'more' than what it now has become, even if this 'more' is not claimed to be 'universal,' or 'biological,' or 'natural.' The contrast alone will not make us change, of course, but it will open the possibility of change.
>
> (Hoy 2004: 63)

The notion of a 'more' that escapes power regimes or history intimates a movement of liberation; however, the 'more' is not necessarily an 'outside' or 'beyond' but an unpredictable manifestation that alters the dynamic of the game. The genealogical or historical perspective of the body opens avenues for critique by employing historical examples of different ways the body has been lived and introducing new norms that disrupt the stability of the norms of the present.

Paul Campos, for example, in *The Obesity Myth* points to West African beauty pageants where 'young women who represent the pinnacle of female beauty in these cultures weigh more than 200 pounds' (2004: 49). Campos argues that not only do different cultures challenge norms of the body and beauty, but also different historical periods within cultures. He says that in regard to desirable body shape, 'contemporary West Africa is quite similar to the United States in the 1890s, when the 200-pound actress Lillian Russell was considered the undisputed beauty of her time' (2004: 49). Campos points to the instability and contingency of current bodily norms in stating that across history 'far more cultures have mirrored contemporary West Africa and America in the latter half of the nineteenth century, than have resembled the United States today, where an almost unprecedented ideal of thinness reigns supreme' (2004: 49–50). The cultural and historical transformations of the body are mobilized as a critical appraisal of present notions of beauty and norms of the body (Gilman 2008, Fraser 2009). This is not to suggest the need to return to an origin or natural bodily form; rather genealogy provides a contrasting perspective to the biopolitical rationality that tries to normalize all bodies – from West Africa to the West of England – to accord with norms based on statistical measurements. These histories serve to demonstrate that the present can be lived differently.

While a genealogical critique can problematize present norms of the body by showing the different historical and cultural ways the body has been lived, empirical and conceptual questions remain regarding biological norms and the idea of fatness. Lillian Russell and the Venus of Willendorf[6] may indicate that other cultures or historical periods valued large bodies, but does this entail that biological norms are plastic or that social norms are not intermingled with them? Is *every* body size really conducive to health, or is there a point when the body size is pathological?

In response to empirical questions about body weight and health, there is a small but growing body of research supporting aspects of the HAES attempt to delink body weight from health (Provencher et al. 2009, Bacon and Aphramor 2011). This research has been used to suggest that a HAES could be incorporated into public health policies to address problems of weight stigma and overemphasis on individual behaviours (Bombak 2013, Penney and Kirk 2015). In addition, Katherine Flegal et al.'s epidemiological research, which still works within a framework that links body weight to health, has broadened the sphere of 'healthy weight' to include overweight (BMI of 25≤30) and to show that obese grade 1 (BMI of 30≤35) is not associated with higher mortality rates (Flegal et al. 2013). However, within these attempts to broaden the scope of normal are explicit or implicit acknowledgments that there are biological limits indicating pathological bodies. For Flegal et al. a BMI >35 is explicitly outside the acceptable range of a normal body. Although Flegal et al. have expanded the range of normal, there is still a statistically determined point when an individual body moves from the normal to the pathological and therefore requires biopolitical governance.

The 'setpoint theory' used by the HAES approach implies that biological norms of weight and body size do not indicate health and are internally regulated. There are similarities between the 'setpoint theory' and Canguilhem's idea that health is the capacity of an organism to produce and pursue new norms of life in its environment. Canguilhem also describes health as 'a margin of tolerance for the inconstancies of the environment' (2007: 197). On this view an individual with a BMI of 37 may have a greater capacity to produce news norms and tolerate environmental shifts than an individual with a BMI of 25. On this view, disease or risk of disease is not related to an individual's size or weight, but the capacity to produce norms. A diseased existence is a 'narrowed mode of life' that cannot tolerate environmental shifts (Canguilhem 2007: 188). The shift in focus from weight to health results in a corresponding shift from biological to social norms of health, which introduces more subtle indicators of pathological or suspect bodies.

Like the public health and biomedical approach to health, the norms of fitness, activity, recreation and diet operating in the HAES movement are also tied to domains of economics, race, education, gender and sexuality. There is an awareness of some of these problems (Burgard 2009: 50–51), however there is also a common perception that there are 'good fatties' and 'bad fatties' (Bias 2014). The goodness or badness of the 'fatty' is dependent on the way certain social norms are negotiated. Writing in *xoJane*, Cary Webb expresses her frustration at the dominance of certain fat acceptance norms – 'eat healthy, exercise regularly…

love your body, resist stigma and shame, wear what you want, date who you like, and only flaunt your sexuality if your body is under a size 20'. Webb notes that despite the emphasis on acceptance there is simultaneous exclusion of certain voices from the size acceptance movement (2015). She writes,

> what alienates me most is the fact that none of those voices look like me and I'm left to wonder, 'Where do I fit in now?' For the record, I'm Black and weigh about 355lbs, have chronic conditions, and while I'm educated, I'm more lower middle class than upper. I'd probably be considered a bad fatty and so would a lot of other fatties I know.
>
> (Webb 2015)

The division between 'bad' and 'good' fatties corresponds to certain social, economic and racial markers. Even with the expanded biological norms via Flegal or the de-emphasis on body weight in HAES, the social norms remain to indicate that certain bodies that are questionable and problematic. I return to these questions in the next chapter.

The persistence of social norms to divide and exclude bodies raises questions about the usefulness of genealogical critique in resisting the biopolitical imperative to adopt a healthy lifestyle. A genealogical critique of bodily norms does not provide an obvious normative position that can resolve disputes over normal and healthy bodies. However, it enables the possibility of thinking differently and becoming 'different from the way we normally are' (Hoy 2007: 3). Hoy argues that it was for this reason that Foucault went back to the Greeks. While Žižek and others believe Foucault's turn to the Greeks was a flight away from the contradictions inherent in his genealogical analyses of power/knowledge (1999: 250ff), others contend this move was a continuation of his interest in the subject and processes of ethical subject formation (O'Leary 2003, Nealon 2008). In *The Use of Pleasure*, Foucault suggests that we can 'learn to what extent the effort to think one's own history can free thought from what it silently thinks, and so enable it to think differently' (Foucault 1992: 9). The process of genealogical comparison opens the possibility for the inscriptions or 'thoughts' of the body to be expressed differently. Thus, the use of genealogy to demonstrate the different lived experiences of the body can serve to critique and resist the lifestyle *dispositif* that forms our subject identities and bodies as 'healthy'.

While HAES and Foucault consider bodies and pleasures to be an important avenue of resistance, and Hoy and Campos emphasize the historical contingencies of bodily norms, critics question the effectiveness of the body as a site for resistance (Fraser 1983). They argue that the subject is subsumed in a body that lacks internal or essential norms, intentionality and agency.[7] Butler suggests that Foucault's body, in responding to power, may 'return in a non-normalizable wildness' (1997: 92). According to Butler, the body emerges for Foucault 'as a way of taking over the theory of agency previously ascribed to the subject' (2004a: 185). Resistance of the subject is no longer located in the psyche of a thinking subject, but in the body. Against this, Butler argues that it is a mistake to lay too strong an emphasis on the capacity for the body to resist. The body, according to Butler is 'not a cauldron of resistance impulse', but a node or point through which power transfers, sometimes in unpredictable ways (2004a: 187). Butler contends that Foucault, and by extension the HAES approach, has not provided the adequate theoretical equipment to support a claim that bodies are able to *resist* power relations. This is a significant criticism, *if* resistance is based solely in the body. However, I contend that a plural approach to resistance is needed. Bodies and pleasures disrupt power transference but they are not the sole drivers of resistant activity. In conjunction with practices of critique and counter-conduct, the unpredictability of bodies can be guided in a more precise manner against governmentality and the norms of behaviour.

### Counter-conduct

In searching for a word to describe resistance to governmentality Foucault considered 'revolt' or the idea of overthrowing power as 'both too precise and too strong' (2007: 200). Foucault wanted a word or concept that could describe the 'diffuse and subdued forms of resistance' that corresponds to the dispersed and capillary-like forms of governmental relations of power (2007: 200). Foucault did have a word mind that 'exactly suited' the forms of resistance he was describing, however, he brusquely exclaims, 'I would rather cut out my tongue than use it' (2007: 200). The word is dissidence. The reason for Foucault's strong aversion is unclear but it is likely due to its Soviet connotations or simply that it was hackneyed[8] – 'After all, who does not have his theory of dissidence today?' (2007: 201). Instead of using dissidence, Foucault offers the more cumbersome but also more analytically astute, 'counter-conduct'.

Foucault outlines a number of examples of counter-conduct from Church history. These include ascetic, interpretive and communal practices that resisted ecclesiastical authorities. The significance of historical examples of counter-conduct is that they did not come from outside the Christian Church 'but are actually border-elements…which have been continually re-utilized, re-implanted, and taken up again in one or another direction' (Foucault 2007: 214–215). Foucault argues that, 'the struggle was not conducted in the form of absolute exteriority, but rather in the form of the permanent use of tactical elements that are pertinent in the anti-pastoral struggle' (2007: 215). Thus, in looking for counter-conducts to the strategies and tactics of governmentality operating in the *dispositif* of lifestyle, it is not necessary to go outside neoliberalism or health promotion. Rather the borders, the knots, clefts and points of contestation within a network can be used to initiate counter-conduct opportunities.

The figure of the pastor and event of the Reformation are central to Foucault's analysis of counter-conduct in *Security, Territory, Population*. An initial problem for Foucault is how to designate the 'specific web of resistance to forms of power that do not exercise sovereignty and do not exploit, but "conduct"' (2007: 200). Foucault raises the problem of designating resistance in relation to a specific form of power: the governmental focus on conduct, or 'acting upon action' (1983: 220).

According to Foucault, 'counter-conduct' appropriately designates the resistance to the forms of power that conduct the 'life and daily existence' of individuals and populations. Counter-conduct references the active sense of resisting. Rather than misconduct, counter-conduct emphasizes the active 'sense of struggle against the processes implemented for conducting others' (Foucault 2007: 201). Counter-conduct places the subject not in opposition, but in a dynamic relation to the governmental grasp that seizes life and directs daily practices.

In 'The Subject and Power' Foucault states, 'the equivocal nature of the term *conduct* is one of the best aids for coming to terms with the specificity of power relations. To "conduct" is at the same time to "lead" others…and a way of behaving within a more or less open field of possibilities' (1983: 220–221). The shifting meanings of conduct highlight the diverse effects of power relations and the need for counter-conduct to be flexible in resisting the effects of power relations. In the context of lifestyle, where the objects (everyday life and bodies) are as diverse as the techniques of control, practices of counter-conduct need to be subtle. Resistance to the network of technologies of power and technologies of the self that conducts conduct provokes a particular response, 'inflaming certain points of the body, certain moments in life, [and] certain types of behaviour' (Foucault 1998: 96). Counter-conduct provides clear utility in charting a critical response to the conduct of conduct at issue in the *dispositif* of lifestyle.

An analysis of counter-conduct to pastoral power makes the avenues and movements of resistance clearer. Foucault states that his 'reason for taking the point of view of pastoral power was, of course, in order to try to find the inner depth and background of the governmentality' that developed in the sixteenth century (2007: 215). The early history of governmentality and counter-conduct strategies employed and enables the possibility for new counter-conducts to respond to contemporary forms of conduct of bodies and lives. Foucault suggests through an analysis of the tactics and strategies of pastoral power and pastoral counter-conducts, that there are:

> 'points of entry' through which processes, conflicts, and transformations… can enter into the field of the exercise of the pastorate, not to be transcribed, translated, and reflected there, but to carry out divisions, valorisations, disqualifications, rehabilitations, and redistributions of every kind.
>
> (Foucault 2007: 216n)

The lesson from the pastorate of the Middle Ages, as the background of governmentality, is that conduct and counter-conduct enter through opportune openings and tactical spaces. These are not stable end points but are re-folded, re-deployed and re-used. The vulnerability of counter-conducts to be re-deployed in new forms of power relations necessitates an attentive attunement to the movements and dynamics of tactical knots, strategic points and openings for counter-conducts to reform. It is here that a third thread of resistance can be added to bodies and counter-conduct.

### *Critique*

In conjunction with the resistant body, critique can provide guidance for counter-conduct. Foucault develops his account of critique via Immanuel Kant and relates it to governmentality. Foucault states that governmentality 'cannot apparently be dissociated from the question "how not to be governed?"' (1997: 44). There is not a simple opposition between governmentality and the governed but a web of discourses, tools and locales through which governmentality operates, provoking multiple and diverse responses. Rather than a blunt revolt directed against a central figure, law or sovereign, Foucault identifies 'a perpetual question which would be: "how not to be governed *like that*, by that, in the name of those principles, with such and such objective in mind and by means of such procedures, not like that, not for that, not by them"' (1997: 44).

The question 'how not to be governed *like that*?' is an important key in examining the possibility of critical resistance in the context of obesity, where everyday practices associated with health and the body are the objects of governmentality. It is here that the significance of Foucault's ethics of the self becomes more pronounced. The critical resistance to governmental conduct is ethical in the sense of producing an ethos or way of life that counters and resists the norms of conduct mobilized through governmentality. Examining the possibility of the individual in the *dispositif* of lifestyle to resist and critique attempts to be 'governed like that, in that way' involves the ethical relation of the subject to itself and others.[9]

To fully grasp the idea of critique and the ethics of the self it is important address Foucault's contested relation to Kant. A number of works explore Foucault's relation to Kant, critique and the Enlightenment (Cutrofello 1994, Allen 2003, Schmidt and Wartenberg 1994, Koopman 2013). However, Foucault's relation to Kant and the Enlightenment is often interpreted via a Nietzschean lens that gives precedence to power while downplaying or ignoring Foucault's claim to be in the tradition of critical philosophy (Foucault 1994: 148). The circulation of Foucault's late essays 'What is Enlightenment?' and 'What is Critique?' and the publication in English of the *Government of the Self and Others 1982–83* course has given greater support to Foucault's discussion of Kantian critique or problematizing the present and ourselves. Foucault introduces the notion of critique as 'the movement by which the subject gives himself the right to question truth on its effects of power and question power on its discourses of truth' (1997: 32). Critique, according to Foucault, enables the possibility for the subject to question the veracity of the effects of power, to 'insure the desubjugation of the subject' (1997: 32). Critique is a process through which the subject can resist subjugation and opens the possibility for 'desubjugation'. This provides a crucial step in the development of a conduct that counters the normalizing effect of lifestyle.

In *The Government of Self and Others*, Foucault examines the problem of the relationship between government of self and government of others. In this problem Foucault recognizes the importance of critique and its relation to the present in Kant's 'What is Enlightenment?' Foucault reads Kant as introducing 'a new way of posing the question of modernity', questions that are 'no longer in a longitudinal relationship to the Ancients, but…a vertical relationship of the discourse to its own present reality' (2010: 13–14). Philosophical discourse could no longer avoid being part of a present 'we' (Foucault 2010: 13). The particular 'we' becomes the object of philosophical reflection and becomes a distinctive feature of the discourse of, and on, modernity. Kant's reflection on the Enlightenment opens up a way of thinking about the present not as a period of history but as an attitude towards the present. Foucault conceives of an attitude as a 'mode of relating to contemporary reality' that is not merely cognitive or doctrinal but a way of feeling, acting and behaving that 'marks a relation of belonging' to a present (2000e: 309). Further, Foucault suggests the 'mode of relating to contemporary reality' is 'a bit like what the Greeks called an *ēthos*' (2000e: 309). Thus the critical task of philosophy joins up with an ethical task of caring for ourselves, such that our mode of relating to the present is expressed through and bound to our bodies, dress, thought, speech and walk. Foucault concludes his own essay 'What is Enlightenment?' by suggesting:

> The critical ontology of ourselves must be considered not, certainly, as a theory…[but] as an attitude, an ēthos, a philosophical life in which the critique of what we are is at one and the same time the historical analysis of the limits imposed on us and an experiment with the possibility of going beyond them.
>
> (Foucault 2000e: 309)

History constitutes the present and permeates daily existence. As such, forming a critical response cannot be a mere theoretical response, but needs to be in the form of an ēthos or attitude that responds to the constituting effects of history.

The critical attitude of the subject is 'the art of not being governed quite so much' (Foucault 1997: 29). According to Foucault, the cultivation of a critical attitude 'is akin to virtue' (1997: 25). Like virtue, critique is an embodied activity through which norms of behaviour can be disrupted, destabilized and transformed (Mills 2010). Butler suggests that the notion of critique as virtue provides a response to accusations from Habermas, Nancy Fraser and others that Foucault's resistance lacks normative direction.[10] Critique-as-virtue, according to Butler, has 'strong normative commitments that appear in forms that would be difficult, if not impossible, to read within the current grammars of normativity' (2004b: 306). By the grammars of normativity, Butler suggests that the normative question of Habermas and critical theory 'what are we to do?' assumes a stable, known and agential 'we'. Foucault's 'we' however, does not allow the question 'what are we to do?' to determine the 'we' as prior. That is, the 'we' is produced through collective practices that can serve to resist effects of governmental strategies aimed at individuals.

In political engagement, Foucault does not appeal to any of those '"we's" whose consensus, whose values, whose traditions constitute the framework for a thought and define the conditions in which it can be validated' (2000d: 114). Unlike Habermas, Foucault does not draw on a predetermined or consensus-based 'we' to provide a normative and legitimate basis for action. Rather, Foucault states '"we" must not be previous to the question; it can only be the result – and the necessarily temporary result – of the question as it is posed in the new terms in which one formulates it' (Foucault 2000d: 114–115). Foucault does not deny the 'we', but regards it as a consequence of critique or activity. The 'we' does not guarantee 'normative commitments' but is produced through a disruptive, transformative and critical ethos. According to Butler, the contribution that Foucault makes to normative theory is strong, but relies on a different grammar that does not accommodate the question 'what are we to do?' without first asking, 'who are we?'

Butler argues that Foucault's ethics of the self establishes a thread between the ethical, aesthetic and political subject. Through work on Greek subject formation, Foucault positions 'his own thought as an example of a non-prescriptive form of moral inquiry'. According to Butler, Foucault is not establishing ethos as 'a *way* of complying with or conforming with preestablished norms', but as 'a critical relation to those norms, one which…takes shape as a specific stylization of morality' (2004b: 308). Foucault's use of Kant to resist authority and develop a critical ethos evidences continuity rather than a break with the Enlightenment project (Butler 2004b: 311). The thread of resistance and critique is the vital legacy of the Enlightenment that necessarily takes on new forms but is still very much relevant to 'our present'.[11]

For Foucault, critique becomes an ethical and political process in response to governmentality and norms of behaviour. Foucault asks: 'if governmentalization is indeed this movement through which individuals are subjugated', which I have argued throughout in relation to lifestyle and obesity, then 'critique will be the art of voluntary insubordination' (Foucault 1997: 32). The process of voluntary insubordination through the body, counter-conduct and critique insures 'the desubjugation of the subject in the context of what we could call…the politics of truth' (Foucault 1997: 32). Thus while the operation of governmentality through the *dispositif* lifestyle is to 'make visible' the healthy subject, the contestation of lines of power and knowledge through critique enables the subject to question norms of behaviour and being 'governed like that'. And in questioning the norms and adopting counter forms of conduct, the possibility of desubjugation is opened, allowing a transformation of the self.

Just as technologies of governmentality attach themselves to the body and to practices of the self, so too does the critical attitude in counter-conduct. In concluding her paper, Butler emphasizes the danger to the subject in adopting a critical attitude to the subjugating effects of governmentality. Through the self-transforming effect of critical practice of the self the subject 'risks its deformation as a subject, occupying that ontologically insecure position which poses the question anew: who will be a subject here and what will count as a life' (Butler 2004b: 321). In entering this process of questioning and critique, the individual is vulnerable to the possibility of deformation as a subject. In resisting the norms at work within the population, the individual exposes themselves to exclusion, isolation and interrogation of 'what kind of subject are you? Does your life count?' To adopt such an attitude in relation to norms of health and expectations of bodies can be a dangerous enterprise. Health is tightly bound to ideas of the health and viability of the group, the population and the polis. To disrupt critique and resist those norms can be perceived as a threat to the whole and justify more targeted interventions.

The line of questioning about who is a subject and whether your life counts enables the critical attitude to inform ethical, political and aesthetic transformation and opens the possibility for the subject 'not to be governed like that, for that, by them'. Critical resistance is not a mere 'desubjugation' risking dissolution, but through the critical attitude the subject enters into a 'we', one constituted in the process of critique. In the 'we', the process of *re*subjectification opens the possibility of critical care for the self and for others, which creates and transforms norms through new practices of freedom. In the following chapter, I address the importance of relations with others in the formation of subjectivity and resisting biopolitical governance of life.

## Notes

1 Bacon is representative of the wider movement, although not synonymous. The critical roots of HAES can be traced through works such as (Orbach 2006); (Gaesser 2002); (Campos 2004); and (Wann 1999).

2 Slavoj Žižek criticizes Foucault on this point, arguing that he does not demonstrate *how* the subject can resist, arguing that 'the very subject who resists these disciplinary measures and tries to elude their grasp is, in his heart of hearts, branded by them, formed by them' (Žižek 1999: 252).

3 'The healthy weight that your body aims for is called your *setpoint weight...* your body's innate weight-regulation mechanism' (Bacon 2010: 13).

4 The relationship between the body and subject in the HAES requires further and lengthier analysis than is necessary here. For instance on this point the 'conflict' in agency between the body and subject suggests an apparent dualism. On the one hand the subject 'chooses' poorly and against the body, but if the subject acquiesces to the body, then the body will choose appropriately and responsibly. Furthermore, the HAES subject is closely related to the neoliberal entrepreneurial subject. I develop some of these points below.

5 Through genealogical analysis Foucault writes 'the body is the surface of the inscription of events (traced by language and dissolved by ideas), the locus of the dissociation of the Me (to which it tries to impart the chimera of a substantial unity), and a volume in perpetual disintegration. Genealogy, as an analysis of descent, is thus situated within the articulation of the body and history. Its task is to expose a body totally imprinted by history and the process of history's destruction of the body' (2000c: 375–376).

6 Linda Bacon argues that '...historians have found numerous depictions of the Venus of Willendorf (dating back to 24,000–22,000 BCE), a stone female figure and a symbol of female beauty. She had a beautiful rounded stomach, big hips, and huge breasts... It wasn't until the 1830s that thinness first came into vogue in North America' (2010: 146). While Lisa Isherwood uses the example of 'Kenyan women who when surveyed did not understand the concept of dieting and...Samoan women who glory in their abundant flesh' (Isherwood 2007: 123).

7 For a survey of the different critical positions see (Oksala 2005).

8 Despite Foucault's instant dismissal of dissidence he does provide a brief account of why it is 'exactly suited' to designate resistance to governmental power that conducts

'life and daily existence'. Firstly, he states that dissidence 'has often been employed to designate religious movements of resistance to pastoral organization' (2007: 200), which as discussed previously is the historical root of governmentality and biopolitics. Secondly, it 'does designate a complex form of resistance and refusal...in a society where political authority... [is] responsible for conducting individuals in their daily life' (2007: 201). Jessica Whyte has a useful discussion of Foucault's hesitation over the word 'dissident' (2014: 223–224).

9   As discussed earlier Foucault conceives governmentality as implying a 'relationship of the self to itself' that covers a 'whole range of practices that constitute, define, organize, and instrumentalize the strategies that individuals in their freedom can use in dealing with each other' (2000a: 300).

10  Nancy Fraser argues that Foucault cannot answer *why* the subject should resist and 'ends up, in effect, inviting questions that it is structurally unequipped to answer' (Fraser 1998: 27). For a similar argument see (Habermas 1987). See also (Taylor 1989: 99).

11  For further discussion of Foucault's idea of enlightenment in relation to ethos see (Osborne 2003: 12).

## References

Allen, Amy. 2003. "Foucault and Enlightenment: a critical reappraisal." *Constellations* 10 (2):180–198.

Bacon, Linda. 2010. *Health at Every Size: The Suprising Truth About Your Weight*. Dallas: Benbella Books.

Bacon, Linda, and Lucy Aphramor. 2011. "Weight science: evaluating the evidence for a paradigm shift." *Nutrition Journal* 10 (1):9.

Bias, Stacy. 2014. "12 good fatty archetypes". http://stacybias.net/2014/06/12-good-fatty-archetypes/

Bombak, Andrea. 2013. "Obesity, health at every size, and public health policy." *American Journal of Public Health* 104 (2):e60–e67. doi: 10.2105/AJPH.2013.301486.

Bordo, Susan. 2003. *Unbearable Weight: Feminism, Western Culture, and the Body*. Berkeley, CA: University of California Press.

Burgard, Deb. 2009. "What is 'healthy at every size'?" In *The Fat Studies Reader*, edited by Esther Rothblum and Sondra Solovay. New York: New York University Press.

Bussolini, Jeffrey. 2010. "What is a dispositive?" *Foucault Studies* 10:85–107.

Butler, Judith. 1997. *The Psychic Life of Power: Theories in Subjection*. Stanford, CA: Stanford University Press.

Butler, Judith. 2004a. "Bodies and power revisited." In *Feminism and the Final Foucault*, edited by Dianna Taylor and Karen Vintges. Urbana, IL: University of Illinois Press.

Butler, Judith. 2004b. "What is critique? An essay on Foucault's virtue." In *The Judith Butler Reader*, edited by Sara Salih and Judith Butler. Malden, MA: Blackwell Publishing.

Callahan, Daniel. 2013. "Obesity: chasing an elusive epidemic." *Hastings Center Report* 43 (1):34–40.

Campos, Paul. 2004. *The Obesity Myth: Why America's Obsession with Weight Is Hazardous to Your Health*. New York: Gotham Books.

Canguilhem, Georges. 2007. *The Normal and the Pathological* Translated by Carolyn R. Fawcett. New York: Zone Books.

Carter, Stacy M., and Helen L. Walls. 2013. "JAMA forum: separating the science and politics of 'obesity'." *Journal of the American Medical Association* 2013 [cited

Feburary 20 2013]. Available from http://newsatjama.jama.com/2013/02/14/jama-forum-separating-the-science-and-politics-of-obesity/.

Cutrofello, Andrew. 1994. *Discipline and Critique: Kant, Poststructuralism, and the Problem of Resistance*. Albany, NY: SUNY Press.

Flegal, Katherine M., Brian K. Kit, Heather Orpana, and Barry I. Graubard. 2013. "Association of all-cause mortality with overweight and obesity using standard body mass index categories: A systematic review and meta-analysis." *JAMA* 309 (1):71–82. doi: 10.1001/jama.2012.113905.

Foucault, Michel. 1980. "Body/power." In *Power/Knowledge: Selected Interviews and Other Writings 1972–1977*, edited by Colin Gordon. New York: Pantheon Books.

Foucault, Michel. 1983. "Afterword: the subject and power." In *Michel Foucault: Beyond Structuralism and Hermeneutics*, edited by Hubert Dreyfus and Paul Rabinow. Chicago, IL: University of Chicago Press.

Foucault, Michel. 1992. *The Use of Pleasure: The History of Sexuality Volume 2*. Translated by Robert Hurley. Harmondsworth: Penguin Books.

Foucault, Michel. 1994. "The art of telling the truth." In *Critique and Power: Recasting the Foucault/Habermas Debate*, edited by Michael Kelly. Cambridge, MA: MIT Press.

Foucault, Michel. 1997. "What is critique?" In *The Politics of Truth*, edited by Sylvère Lotringer and Lysa Hochroth. New York: Semiotext(e).

Foucault, Michel. 1998. *The Will to Knowledge: The History of Sexuality Volume 1*. Translated by Robert Hurley. Harmondsworth: Penguin Books.

Foucault, Michel. 2000a. "The ethics of the concern for the self as a practice of freedom." In *Ethics: Subjectivity and Truth*, edited by Paul Rabinow. London: Penguin.

Foucault, Michel. 2000b. "For an Ethics of Discomfort." In *Power: Essential Works of Foucault 1954–1984*, edited by James D. Faubion. Harmondsworth: Penguin.

Foucault, Michel. 2000c. "Nietzsche, genealogy, history." In *Aesthetics, Method, and Epistemology*, edited by James Faubion. London: Penguin Books.

Foucault, Michel. 2000d. "Polemics, politics, and problematizations." In *Ethics: Subjectivity and Truth*, edited by Paul Rabinow. London: Penguin.

Foucault, Michel. 2000e. "What is Enlightenment?" In *Ethics: Subjectivity and Truth*, edited by Paul Rabinow. London: Penguin.

Foucault, Michel. 2007. *Security, Territory, Population: Lectures at the Collège de France 1977–78*. Translated by Graham Burchell. Edited by Arnold I. Davidson. New York: Palgrave Macmillan.

Foucault, Michel. 2010. *The Government of Self and Others: Lectures at the Collège de France: 1982–1983*. Translated by Graham Burchell. Edited by Arnold I. Davidson, Basingstoke: Palgrave Macmillan.

Fraser, Laura. 2009. "The inner corset: a brief history of fat in the United States." In *The Fat Studies Reader*, edited by Esther Rothblum and Sondra Solovay. New York: New York University Press.

Fraser, Nancy. 1981. "Foucault on modern power: Empirical insights and normative confusions." *Praxis International* (3):272–287.

Fraser, Nancy. 1983. "Foucault's body-language: a post-humanist political rhetoric?" *Salmagundi* 61:55–70.

Fraser, Nancy. 1998. "Foucault on modern power: empirical insights and normative confusions." In *Unruly Practices: Power, Discourse and Gender in Contemporary Social Theory*. Minneapolis, MN: University of Minnesota Press.

Gaesser, Glenn. 2009. "Is 'permanent weight loss' an oxymoron? the statistics on weight loss and the national weight control registry." In *The Fat Studies Reader*, edited by Sandra Solovay and Esther D. Rothblum. New York: New York University Press.

Gaesser, Glenn A. 2002. *Big Fat Lies: The Truth about Your Weight and Your Health*. Carlsbad, CA: Gürze Books.

Gilman, Sander. 2008. *Fat: A Cultural History of Obesity*. Cambridge: Polity Press.

Goldberg, Daniel S., and Rebecca M. Puhl. 2013. "Obesity stigma: a failed and ethically dubious strategy." *Hastings Center Report* 43 (3):5–6.

Habermas, Jürgen. 1987. *The Philosophical Discourse of Modernity: Twelve Lectures.* Translated by Frederick Lawrence. Cambridge, MA: MIT Press.

Hoy, David Couzens 2004. *Critical Resistance: From Poststructuralism to Post-Critique*. Cambridge, MA: MIT Press.

Hoy, David Couzens 2007. "Response – reflections on critical resistance." *Foucault Studies* 3(November), 101–106.

Isherwood, Lisa. 2007. *Fat Jesus*. London: Darton Longman & Todd.

Koopman, Colin. 2013. *Genealogy As Critique : Foucault and the Problems of Modernity, American Philosophy*. Bloomington, IN: Indiana University Press.

McWhorter, Ladelle. 1999. *Bodies and Pleasures: Foucault and the Politics of Sexual Normalization*. Bloomington, IN: Indiana University Press.

Mills, Catherine. 2010. "A manner of speaking: declaration, critique and the trope of interrogation." *Law and Critique* 21 (3):247–260. doi: 10.1007/s10978-010-9073-y.

Nealon, Jeffrey T. 2008. *Foucault Beyond Foucault: Power and Its Intensifications since 1984*. Stanford, CA: Stanford University Press.

O'Hara, Lily, and Jane Gregg. 2006. "The war on obesity: a social determinant of health." *Health Promotion Journal of Australia* 17 (3):260–263

O'Leary, Timothy. 2003. *Foucault and the Art of Ethics*. London: Continuum.

Oksala, Johanna. 2005. *Foucault on Freedom.* Cambridge: Cambridge University Press.

Orbach, Susie. 2006. *Fat Is a Feminist Issue*. London: Arrow.

Osborne, Thomas. 2003. "What is a problem?" *History of the Human Sciences* 16 (4):1–17. doi: 10.1177/0952695103164001.

Penney, Tarra L., and Sara F.L. Kirk. 2015. "The health at every size paradigm and obesity: missing empirical evidence may help push the reframing obesity debate forward." *American Journal of Public Health* 105 (5):e38–e42. doi: 10.2105/AJPH.2015.302552.

Provencher, Véronique, Catherine Bégin, Angelo Tremblay, Lyne Mongeau, Louise Corneau, Sylvie Dodin, Sonia Boivin, and Simone Lemieux. 2009. "Health-at-every-size and eating behaviors: 1-year follow-up results of a size acceptance intervention." *Journal of the American Dietetic Association* 109 (11):1854–1861. doi: http://dx.doi.org/10.1016/j.jada.2009.08.017.

Puhl, Rebecca M., and Chelsea A. Heuer. 2010. "Obesity stigma: important considerations for public health." *American Journal of Public Health* 100 (6):1019–1028.

Puhl, Rebecca M., Jamie Lee Peterson, and Joerg Luedicke. 2013. "Weight-based victimization: Bullying experiences of weight loss treatment–seeking youth." *Pediatrics* 131 (1):e1–e9.

Schmidt, James, and Thomas E. Wartenberg. 1994. "Foucault's Enlightenment: critique, revolution, and the fashioning of the self." In *Critique and Power: Recasting the Foucault/Habermas Debate*, edited by Michael Kelly. Cambridge, MA: MIT Press.

Taylor, Charles. 1989. *Sources of the Self: The Making of the Modern Identity*. Cambridge, MA: Harvard University Press.

Wann, Marilyn. 1999. *Fat! So? Because You Don't Have to Apologize for Your Size!*. Berkeley, CA: Ten Speed Press.

Webb, Cary. 2015. "Why I'm over the size acceptance movement or Hey, SA, What have you done for me lately?" *xoJane*, http://www.xojane.com/issues/why-im-over-the-size-acceptance-movement-or-hey-sa-what-have-you-done-for-me-lately.

Whyte, Jessica. 2014. "Is revolution desirable?: Michel Foucault on revolution, neoliberalism and rights." In *Re-reading Foucault: On Law, Power and Rights*, edited by Ben Golder. New York: Routledge.

Žižek, Slavoj. 1999. *The Ticklish Subject: The Absent Centre of Political Ontology*. London: Verso.

# 7 Relations of care

## Restless and endless transformation

Riots, not diets!
Fat acceptance and feminist slogan

The neoliberal rationality of governance focuses on the individual as responsible for health, education and all manner of life events, regardless of circumstance or history. To counter this rationality we need to move beyond this methodological individualism that understands the social in terms of the individual. This concept was introduced in Chapter 1 to highlight the way Peter Singer and Dan Callahan justified their bioethical arguments that individuals who are obese or engaged in practices associated with being obese should be coerced and stigmatized. Methodological individualism not only frames the type of governmental interventions available, but it also sets the limits for the type of critique and resistance available. However, as has been shown through this analysis, the individual is produced through a network of relations. To suggest that more individual freedom or less government intervention will serve to resist these relations misunderstands the problem. This final chapter argues that resistance needs to come from individuals in relations of care with others, which can establish new modes of living together.

As mentioned in previous chapters, Foucault's analyses of ancient practices of the self have been criticized for embracing a self-obsessed ethics and/or for retreating into the ancient world away from the political struggles of modernity (McNay 1992, Žižek 1999, Rochlitz 1992). This chapter argues that the practices of the self are deeply inter-subjective and can provide important clues for collective resistance to the biopolitical mechanisms of governance that focus on individual choices, behaviours and bodies.

Foucault's interest in the relation between practices of the self and others is evident in the lecture series, *The Hermeneutics of the Subject*. Foucault examines Greek and early Christian pedagogical texts to demonstrate the relation between 'governing others' and 'being governed' in the context of caring for the self. Foucault writes, '"Governing," "being governed," and "taking care of the self" form a sequence, a series, whose long and complex history extends up to the establishment of pastoral power in the Christian Church in the third and fourth centuries' (2005: 45). The links between care of the self and pastoral power in practices of subject formation is underdeveloped in the literature on

Foucault. However, pastoral power is a significant theme of biopolitics as an art of government and forms an important relation in resistance to biopolitical governance. Although ordinarily considered as disciplinary, the pastoral relation is volatile and can be redeployed as an ethical relation of care. This relation provides opportunities for the individual to actively participate in subject formation with a 'master' that resists the imposition of norms. The interpretation of pastoral power as a disciplinary relationship emphasizes the production of the individual as a governable subject. This is placed in contrast to the 'care of the self', which establishes a relation of ethics, freedom and subjectification. However, I contend that the relationships established in pastoral power and care of the self share a number of similar and at times indistinguishable characteristics.

Like the duck–rabbit illusion, pastoral power and care of the self can appear to subjugate and to de-subjugate, to violate and to care (Grimshaw 1993). These logics operate through the lifestyle network to produce subjects that are responsible and therefore deserving of care and those that are irresponsible and therefore undeserving. However, the entanglement of lines of power, knowledge and subjectification in this network make it very difficult for subjects to resist and critically question the rationalities by which they are governed.

In proposing strategies of resistance, a common question is how do you measure or even know if resistance is successful? If 'care of the self' mobilizes a critical attitude that allows for desubjugation, on what scale and by what measure can it be deemed successful? Foucault uses both the Reformation (2007: 149ff) and the French Revolution (2010: 17ff) as illustrations of the movements of counter-conduct and critique, but it is doubtful that these are to serve as points against which the success of resistance is to be measured. In addressing the question of what successful resistance looks like, the process and form of resistance becomes more apparent. Resistance is not revolutionary, it is not permanent and it is not individual; or in the positive form, resistance is everyday, restless and communal. Applying these models to the lifestyle governance of healthy choices and bodies, these three features can guide and enable the individual to contest subjection according to public health and economic norms, and transform the self through relations of care. The combination of these features opens the possibility for the individual to resist biopolitical governance of the body and everyday conduct in the *dispositif*.

## Everyday resistance

Earlier, I discussed the turn to the everyday through consumption, epidemiology and economics that mobilize a network of lifestyle power/knowledge. In response to the 'urgent need' of obesity, the everyday has become a contested field in which the techniques of discipline and the self are vulnerable to redeployment to create paths of resistance. It necessary to address the role of resistance at the level of the everyday because it is through shaping the daily choices, habits and activities that technologies of governmentality operate. A further reason why Foucault rejected the word dissidence to describe resistance is that it connotes 'a process of sanctification or hero worship' (2004: 202). Heroic resistance is rejected for two reasons. First, it is not effective due to the tactical and porous effect of governmentality that targets the life and daily existence of the subject. The neoliberal rationality of governance is too flexible to be vulnerable to a macro-resistance coming from a central force, be it 'the people', 'the workers' or 'the students'. De-centralized power emanates from a plurality of locales, particularly in mundane consumer choices and everyday activities, meaning that resistance needs to be of a similar form to counter the power effect. For example, a dictator who centralizes power in his position is vulnerable to a popular uprising. If the people overthrow or kill the dictator then everything changes, for good or ill. In neoliberal consumerist society, however, the operation of power is through micro-practices and therefore requires a micro-resistance.

The second reason why heroic resistance should be rejected is that it suggests single identities are the responsible and active agents of social change. Focusing on single heroic actors as the cause of social upheaval and change does not account for the variety of conditions required for such change to be possible. Feminist protest and collective action has significantly shaped debates. Calls for 'riots not diets' are symbolically and rhetorically powerful (Webb, 2015: 1787). However, in the context of obesity governance, the norms of health and the body are created and disseminated from a plurality of locales: celebrities and health ministers, academic researchers and app designers, lifestyle journalists and physicians. A single revolt cannot overthrow the influence of these norms in moulding the behaviour and bodies of individuals toward particular ends or goods that the subject does not necessarily adhere to.

The lifestyle network enables biopolitical governance to direct the individual to adopt healthy choices that promotes their own health and secures the population from the perceived threat of obesity. This objective requires a variety of biopolitical techniques to regulate and discipline irregularities in the everyday life of the individual. Resistance to these techniques and imperatives do not take the form of a heroic resistance that would overthrow health promotion, neoliberalism or lifestyle media. Rather, the subject needs to stage irregular forms of resistance in the 'everyday struggles with power' at the level of everyday existence (Nealon 2008: 111).

As discussed in the previous chapter, an example of this form of resistance is seen in the Health At Every Size (HAES) emphasis on health rather than body size or shape. The HAES approach 'aims to reduce barriers to participation in nutrition, physical activity, positive body esteem, social support, and community connection activities for people of all sizes' (Lyons 2009: 84). Resistance in this form is not in direct opposition to the health promotion strategies of *Measure Up* or *Let's Move* but it does undermine the emphasis on waistline and weight. Tactical and strategic shifts force the game in a new direction where health is still the goal but weight and body shape is no longer the target.

The HAES approach attempts not only to counter the weight–health nexus of obesity governance, but also to challenge the weight–aesthetic nexus established through the prescription of micro-practices in lifestyle media, smartphone apps and expert guidance. The 'good', 'healthy' and 'beautiful' body is not necessarily lived between a BMI 18.5 and 24.9 or waistline under 94 cm for men and 80 cm for women. By focusing 'on the day-to-day activities', HAES seeks to 'help individuals of any size to flourish' (Burgard 2009: 43). In emphasising 'self-acceptance and healthy day-to-day practices' (Burgard 2009: 42), HAES tries to 'reclaim the worth of our stigmatized bodies and encourages subversive acts of self-care' (Burgard 2009: 52). The success of HAES in contesting the micro-practices of the everyday will be dependent on social context and community. However, what is evident is that the contestation over everyday conduct can be redeployed through tactical games of strategy. The HAES 'solution' is by no means final or exhaustive, the resistance of the body and techniques of counter-conduct are able to force the lines of power operating through the lifestyle *dispositif* to retreat, but only to 're-organise its forces, invest itself elsewhere…and so the battle continues' (Foucault 1980: 56).

## Restless resistance

The agonistic logic, or what Foucault describes as the continuing battle and re-organization of forces, is an important feature of resistance to norms of behaviour. In addressing the normative question, 'what makes a "good" or "successful" form of resistance?', Foucault states that he is not interested in prescribing the right course of action, denouncing one form of power as illegitimate and another as legitimate or mapping the path to the subject's complete emancipation. He insists that:

> My point is not that everything is bad, but that everything is dangerous, which is not exactly the same as bad. If everything is dangerous, then we always have something to do. So my position leads not to apathy but to hyper- and pessimistic activism.
>
> (Foucault 1983: 231–232)

In stating that 'everything is dangerous' Foucault commits himself to a hyperactive vigilance toward everything yet is pessimistic regarding a final resolution. The pessimism of a resolution is not the failure of the particular strategy of resistance or counter-conduct. Rather, success is simply not measured by the achievement of a final goal at which one can rest. The need for hyper- and pessimistic activism is due to the flux of problems, leading to an endless resistance. Endless here implies both continual and without a specific *telos*.

Refusing to offer a normative set of criteria that would determine successful resistance frustrates critics such as Habermas. The agonistic logic of Foucault's answer means that *Measure Up* is 'bad' and HAES is 'good', but both are dangerous. The HAES critique of dieting and norms of healthy lifestyle regimens is 'dangerous'. As discussed in the previous chapter, these communities can reinstate new norms, such as 'good fatties' and 'bad fatties', which limit and exclude certain lives. It is in this sense that relations of care can also result in exclusion (Webb 2015). Furthermore, HAES is dangerous not only for the reinstatement of new norms of living or for its threat to the diet industry but also in the way it can be co-opted and redeployed through the enabling network of lifestyle. Bacon asks her readers to 'face it, the "D" word is dead. A new diet isn't going to get you what you want' (2010: 1). However, a number of lifestyle authorities also admit the 'diet is dead'. An issue of *Women's Health*, a lifestyle magazine promoting health and beauty, reaches the same conclusion, stating, 'by some estimates…80 per cent of people who've lost weight regain it all, or more, after two years' (Voss 2010: 105). Like HAES, the article acknowledges that willpower and laziness are not the reason behind the failure of dieting. Mirroring the HAES approach, the article refers to the body as 'hardwired' to a natural 'setpoint' and employs the 'thrifty gene hypothesis'[1] to explain why the body resists dieting. However, instead of concluding that dieting is futile and body weight benign, the article argues for greater vigilance across a lifetime. Individuals are admonished to develop 'the skills needed for long-term behaviour change' (Voss 2010: 106), which can be developed via behaviour change therapy in addition to weight-loss techniques such as recording 'mood changes and hunger levels', to vigilantly 'weigh yourself weekly' and to not 'think of what you're doing as "dieting"… [but] as a permanent shift: "This is how I eat now"' (Voss 2010: 106). The critique of dieting and the diet industry is vulnerable to redeployment as supporting evidence for a stronger and more aggressive application of biopolitical norms in the need for a 'permanent shift' employing behavioural change therapy.

A more drastic or 'dangerous' redeployment of the critique of dieting is seen in the use of bariatric surgery as a mechanism to develop a healthy body and secure the threat of obesity. Writing in the *Medical Journal of Australia*, O'Brien, Brown and Dixon assert that obesity is 'one of the greatest health challenges of the 21st century in Western countries' (2005: 311). Yet, they argue that this challenge is not met through dieting. Drawing a similar conclusion to the HAES approach, O'Brien, Brown and Dixon state, 'solutions that involve lifestyle change are simple to prescribe, yet rarely achieve sustainable outcomes' (2005: 312). Acknowledging the gap between lifestyle rhetoric and practice, they argue that 'bariatric surgery is the only current treatment that has been shown to achieve major and durable weight loss' (2005: 310). Thus the critique of lifestyle techniques of bodily discipline can lead toward the fat acceptance and HAES movement, but it can also expose bodies and subjects outside of the norm to more aggressive and violent techniques. Bariatric surgery has been described as a form of mutilation (Royce 2009: 155), a betrayal of self and others (Bernstein and St. John 2009: 267), stomach-amputation (Wann 2009: xxi), and a traumatic life-threatening experience (Murray 2008: 166). According to Bacon, bariatric surgery is an irreversible and 'high-risk disease-inducing cosmetic surgery' (2010: 62). The mobilization of the critique of diet to justify bariatric surgery highlights the vulnerability and contested nature of knowledge in the lifestyle network of governance, which underscores Foucault's suggestion that his position leads to 'hyper- and pessimistic activism' (1983: 232).

Critique opens spaces for new practices, bodies and pleasures. Yet these spaces are not stable or permanent, but continually contested. Foucault maintains that we are not trapped by power, politics or subjectivities, but 'there are always possibilities of changing the situation' (2000c: 167). The possibility of change does not open out into a space beyond the struggle but to temporary reprieves and new struggles. And it is in this endless possibility of struggle that new pleasures of norms and subjection are fostered. The inclusion of pleasure and self-creation as part of resistance has attracted criticism from McNay, who has characterized this aspect of Foucault's work as frivolous and 'unregulated libertarianism' (1994: 159). These charges mirror the accusations levelled at youth movements like Occupy Wall Street and the Hippies before them, which started out as counter-cultural but were soon absorbed into the mainstream consumer culture (Binkley 2007). However, if the agonism of Foucault's resistance is maintained, if the hyper- and pessimistic activity of restless and endless critique is fostered, then openings can be created that desubjugate and enable possibilities for re-subjectification, if only for a time.[2] Importantly, the process of re-subjectification is not performed in self-imposed isolation or exile but in the context of relations with others. It is not individualistic, but is fostered through the ethics of the self in community. It is through resistance with others that the charge of 'unregulated libertarianism' can be avoided.

## Resisting with others

Relations with others can open new possibilities for desubjugation and resistance. Relationships with others serve as space for mutual subject formation and production of a new 'we'. Although aligned with notions of community, I do not intend to theorize relations with others in the context of community in Foucault's work. Community is a significant theme in biopolitical theory with a number of important works emphasizing its potential for political resistance, new forms of life or global political order (Agamben 1993, Esposito 2009, Olssen 2010, Hardt and Negri 2001). Rather I focus on relations with others in the context of the relationship between a master or pastor figure and an individual. These relations occur through loose social interactions among individuals attempting to live a similar way of life, or lifestyle, together. I argue that the care of the self relation incorporates both of these relational aspects and allows the individual to develop a critical ethics of the self that resists subjugation through the lifestyle *dispositif*.

The care or ethics of the self is described by Foucault as an attitude, stance or position of the self toward the self – a call to 'Attend to your self' and for individuals to 'be concerned about themselves' (2005: 52–57, 6). The attention of the self to oneself is guided through a series of practices such as meditation, memorization, self-examination, withdrawal, self-writing, dietetics and confession (Foucault 2005: 11, 59, 1990: 50–51). The care of the self, through a diversity of practices of freedom, enables the individual to govern themself and others, developing and forming the self as an ethical subject. Yet it is also a social practice that requires guidance from a 'master'.

In outlining the importance of a master, Foucault employs similar terms as when describing the confession in pastoral power. In *The Hermeneutics of the Subject* Foucault states, 'there is no care of the self without the presence of a master' (2005: 58). Akin to the presence of the pastor in the confession, 'the care of the self is actually something that always has to go through the relationship to someone else who is the master' (Foucault 2005: 58). In 'The Ethics of the Concern for Self as a Practice of Freedom', Foucault re-emphasizes the importance of relationship, 'the care of the self also implies a relationship with the other insofar as proper care of the self requires listening to the lessons of a master' (2000a: 287). Thus not only does the care of the self employ similar techniques to pastoral power (self-monitoring, self-examination and confession), it also requires the presence of a master or guide to lead the individual in the formation of the ethical subject. This master could be a health professional, counsellor, priest, teacher, mentor or friend.

A complex and important entanglement unfolds between biopower and care of the self. Pastoral power is the pivot for this entanglement. The role of the master in caring and enabling self-care is performed in both pastoral power and the care of the self. This is not to suggest that pastoral power and care of the self are identical. The role of master as a guide in formation of the subject provides an important, if uncanny, point of harmony. A certain volatility and danger surrounds the care of the self relation as it can turn into a pastoral relation, and vice versa.

A consequence of this entanglement is that the self-formation involved in the care of the self relationship opens up the possibility of altering the power of the pastoral relationship in producing docile bodies and particular subjectivities. Roberto Esposito points to this possibility in writing that the pastoral relationship can 'support power and increase it, but also…resist power and oppose it' (2008: 38). In the context of lifestyle, the individual alone cannot resist governmentality and the objectives of security. However, with others, the individual can wage resistance against specific tactics and goals. The pastor or master relation is also strategic in critiquing and resisting the norms of behaviour that biopolitical arts of government seek to produce. The care of the self is not a solitary process but an 'intensification of social relations' (Foucault 1990: 53). It is in the context of relations with others that the self is cared for, guided and counselled. Thus an individual could establish a relation of care with a health professional in which the norms of the body in the lifestyle setup are countered and resisted.

Pastoral power as a 'power of care' in concert with the ethical relationship of the care of the self provides an avenue for subject formation in resistance to particular forms of subjectivity generated through biopower. Cressida Heyes notes the ambiguity between pastoral power and the care of the self in her analysis of commercial dieting programmes. Heyes argues, 'the language of commercial weight loss resonates eerily with [Foucault's] own suggestions…about practical training of the self in aid of developing an art of living' (2007: 63–64). Weight Watchers and other dieting programmes cannot be reduced solely to disciplinary practices producing docile bodies but need to be supplemented with Foucault's ethics of the self. Heyes argues, 'Weight-loss dieting needs to be understood from within the minutiae of its practice, its everyday tropes and demands, its compulsions and liberations; and, in turn, these cannot be resisted solely through refusal' (2007: 64).

The function of resistance is problematized in dieting programmes such as Weight Watchers. Unlike the disciplinary techniques used in schools, prisons or asylums, the clients of dieting programmes, users of smartphone apps, and readers of *Prevention* are 'willing participants in a disciplinary technology' (Heyes 2007: 75). The individuals engaging with these practices of the self do so in a bid to create a healthy lifestyle that prevents disease. In addition to willingly entering such programmes, Heyes notes the pleasure and empowerment these programmes give. Based on her experience of Weight Watchers, Heyes writes, 'I began to understand the satisfaction many women found not only in losing weight, but in working on themselves' (2007: 78). It is not simply the production of a normalized body, but the 'active, creative sense of self-development, mastery, expertise, and skill that dieting can offer' (Heyes 2007: 78). These 'skills' extend beyond weight-loss to personal relations, marriage relations and career direction (Heyes 2007: 83).

Weight Watchers, the Eatery or *Prevention* are not bad, but they are dangerous. These practices have the potential for present avenues of resistance and counter-conduct. John Ransom argues that a relationship with a master or pastor figure allows 'the possibility that an individual might be produced [that] is more aware of the possible effects of disciplinary procedures and so stands in a better position to resist them' (1997: 138). Thus a HAES physician or Weight Watchers 'group leader' could serve as pastor or master to an individual seeking to resist lifestyle techniques of weight-loss and body control (Heyes 2007: 65). Through social relationships with a master or pastor – a social relationship – the individual can stage an irregular resistance involved in the 'everyday struggles with power' at the level of everyday life and subjectivity (Nealon 2008: 111). These practices and relations with others open up opportunities to 'create ourselves as a work of art' that are aesthetic and political (Foucault 1983: 237). Although Foucault's emphasis on self-creation has attracted the criticism that he 'privileges the individual's relation with the self over relations with others' (McNay 1994: 152, Grimshaw 1993), I contend that this criticism misreads the role of relationship with others in the care of the self.

Ethics of the self involves a relation with the self *through* relations with others. In situating the ethics of the self in relation with others, it is possible to instigate a critical re-subjectification that resists the biopolitical *dispositif* of lifestyle. The rhetoric of lifestyle media and empowerment discourse of health promotion blurs with the language of the care of the self in ways that would be unrecognizable if they were analysed as mere disciplinary techniques to normalize bodies (Heyes 2006: 144). Like the rhetoric of *Prevention*, Heyes observes that the language of Weight Watchers mirrors the care of the self to 'deepen its members' dependence on the organization' (2006: 140), forming and cultivating the individual as a particular subject embedded in a community. Rather than reducing the attraction to such programmes as false consciousness or alienation, the communal form of these practices and relations has the potential to cultivate the self. Heyes argues that for a strategy of resistance to be successful it needs to offer a network of support equivalent to those offered through diet programmes (2007: 67). Heyes reflects that through Weight Watchers, she saw how 'communities of women could be mobilized (both in face-to-face meetings and online) in ways that beg to be imitated by a diet-resisting not-for-profit feminist organization' (2007: 88).[3] The focus on others in The Eatery, the empowering rhetoric of Oliver and Obama, and the language of Weight Watchers draws attention to the role of community and the ambiguity between technologies of power and technologies of the self, between subjugation and de-subjugation, between violence and care. The ambiguity between dieting regimens, which have traditionally been critiqued as disciplinary, and practices of the self, which are generally regarded as aesthetic and freeing, means that strategies of resistance in these relations become complex and multifarious.

The role of others helps to provide some clarity about practices of resistance. The mobilization of communities that employ techniques of care of the self to resist the bodily norms of the lifestyle *dispositif* is demonstrated in the Health at Every Size and Fat Acceptance movements. In different ways these communities provides avenues through which individuals in relation with others can care for the self. An example from the Fat Studies literature is the Body Image Network (BIN),[4] which draws on a variety of pastors, such as 'sociologists, psychologists, dieticians, kinesiologists, nurses, epidemiologists, physicians, educators and students' (Beausoleil 2009: 96), to critique and resist norms of the body operating in anti-obesity strategies. Natalie Beausoleil describes BIN as engaging youth, community leaders, policy makers and health care professionals 'to help create a social environment that supports resilience to unrealistic expectations of the body' (2009: 96). While Beausoleil suggests this may be an 'impossible task', her critical assessment of BIN's attempt to counteract the influence of biopolitical techniques targeting the obese bodies of young people highlights the importance of pastors and community in establishing resistance. In addition, these relations and communities may not only serve to resist certain strategies and norms, but also model new ways of life that disrupt the circulation of norms in society. Foucault describes a similar process in relation to the creation of the 'homosexual mode of life' and the communities that facilitate that life (2000b: 137). Foucault states:

> A way of life can be shared among individuals of different age, status, and social activity. It can yield intense relations not resembling those that are institutionalized. It seems to me that a way of life can yield a culture and an ethics. To be 'gay', I think, is not to identify with the psychological traits and the visible masks of the homosexual but to try to define and develop a way of life.
>
> (Foucault 2000b: 138)

Friendship or intense relations external to institutions such as marriage, the family, clinic, school or workplace provide opportunities for new forms of life shared among individuals. Foucault recognizes this potential for new ways of living together to produce ethics and culture in gay communities not because of 'the intrinsic qualities of the homosexual' but through the 'slantwise' or 'diagonal lines' that these new forms of living and relating 'lay out in the social fabric' (2000b: 138).

Foucault states the 'homosexual mode of life, much more than the sexual act itself', 'disturbs' social relations (2000b: 136). Foucault's use of the word 'disturbs' can be taken as meaning provocation and disruption, both of which can unpick threads in the social fabric. The 'formless' and non-institutionalized friendship between two men 'short-circuit [institutional codes] and introduce love where there's supposed to be only law, rule, or habit' (Foucault 2000b: 137). In a similar way, fat bodies and communities disturb social relations not only through the mass of the body, but the mode of life established through the community. Rather than being ashamed of a fat body and seeking to conform to medical, moral and aesthetic law, rule, or habit, people of different sizes are able to live 'full, happy, fulfilled lives, and are in satisfying relationships' (Bacon 2010: 191). Establishing a mode of life through relations with others and engaging in processes of desubjugation, the individual can critically resist subjection as diseased or obese. Importantly, Butler's warning that subjects risk 'deformation' by occupying an 'ontologically insecure position' is reduced (2004: 321), but not eradicated, through relations of care with both masters and friends that establish new modes of living.

In a variety of ways, the communities associated with Fat Studies have consciously modelled themselves on Gay and Queer activism of the late-twentieth century. An example of this is the emphasis on fat 'as a preferred term of political identity' (Wann 2009: xii) that rejects the medical descriptors 'overweight' or 'obese'. Loosely composed of institutionalized (e.g. National Association to Advance Fat Acceptance) and non-institutionalized relations, the fat acceptance communities challenge, resist and transform norms of health, aesthetics and the body. Through performance, fiction, fat positive art, blogs, online forums, and fashion, a mode of life is established that creates a shared culture and ethics. For example, there are a huge variety of blogs dedicated to a Fat mode of living. For example, the *Fat Girls Guide to Living* is an interesting inversion of lifestyle media as it tries to provide 'practical advice and resources for women sizes 14+ up as we navigate the often tricky balance of being an overweight woman and living a rich, rewarding life full of all the things we'd love to do today' (Fat Girls Guide to Living 2009).[5] And the 'Fat Studies Yahoo! Group' is an important online community where academic, activist, medical and pastoral resources are shared and discussed. These communal modes of life and bodies are 'slantwise' and anarchic disruptions that alter the social fabric (Oksala 2005).

These relations do not simply desubjugate but offer avenues for new processes of subject formation. Hoy argues that through relations with others desubjugation does not involve people discovering 'who they really are and what they ought to do. Instead…critique challenges their understanding of who they are, and it leads them to resist their attachment to their social identities and ideals' (2004: 14). For this reason, I contend that the suggestion of *de*subjugation of the subject is perhaps misleading as it implies a neutral position where the subject is somehow stripped of identity and freed from being a subject. According to Hoy, 'the point of critique is to enhance the lives and the possibilities of the individuals, to allow them the space to try to create themselves as works of art' (2004: 92). Thus desubjugation through the critical attitude provides the space for *re*subjectification and self-making, and allows for the possibility of a space in which resistance can be said to 'succeed', if only temporarily.

The critical ethics of the self outlined above contributes to the subject's re-formation not as an individual process but through relationships. As demonstrated through examples of HAES and fat acceptance movements, critical resistance promotes a 'social *déassujettissement* [that] could well open the door to social change' (Hoy 2007: 105). The ethics of the self is not destined to be an 'unregulated libertarianism' (McNay 1994: 159) or anarchic self-creation, but through critical community with others new possibilities of an ethics of how 'we' live are opened (Hoy 2007: 105). The 'we' proceeds from critical relations with others. Through relationship with others, the individual can care for their self in a way that enables conversion of 'limiting and controlling power' (Foucault 2000a: 288). Thus in relationship with a pastor or friendships engaged in developing a critical ethos, individuals resist the controlling and limiting effects of the lifestyle network that operates through epidemiological knowledge, neoliberal politics and consumer seduction that define bodies as diseased and direct individuals to make health-promoting choices purporting to secure the population from obesity.

## Notes

1  This hypothesis suggests that genes which store energy in the body as fat were at some point in history beneficial for survival, however in the current context these genes have become a liability for health and survival. For an example of the 'thrifty gene hypothesis' see (Bacon 2010: 23–25).

2  The idea of restless and endless self-formation is evident in Nietzsche's critique of 'enduring habits' in *The Gay Science*. Nietzsche argues that habits can suffocate life but life devoid of them would be intolerable. Nietzsche writes '*[e]nduring* habits...I hate, and feel as if a tyrant has come near me and the air around me is *thickening* when events take a shape that seems inevitably to produce enduring habits' (Nietzsche 2008: 168). However, he concludes that 'the most intolerable, the truly terrible, would of course be a life entirely without habits, a life that continually demanded improvisation – that would be my exile and my Siberia' (Nietzsche 2008: 168).

3  For additional feminist perspectives on ideas of relational care and resistance see (Grimshaw 1993), (Allen 2008), and (McWhorter 1999).

4  'The Body Image Network is a group of individuals and organizations committed to promoting a positive social environment through sharing information on body image, self-esteem, obesity and eating disorders. We are academics, advocates, researchers, dieticians, social workers, counselors, psychologists, doctors, teachers, nurses and students.' http://www.bodyimagenetwork.ca/

5  For examples of Fat Fiction see (Jarrell and Sukrungruang 2003) and (Koppelman 2003). For examples of Fat Fashion see http://www.australianfatshion.com/ and (Garner 2010).

# References

Agamben, Giorgio. 1993. *The Coming Community*. Minneapolis, MN: University of Minnesota Press.

Allen, Amy. 2008. *The Politics of Our Selves: Power, Autonomy, and Gender in Contemporary Critical Theory*. New York: Columbia University Press.

Bacon, Linda. 2010. *Health at Every Size: The Suprising Truth About Your Weight*. Dallas, TX: Benbella Books.

Beausoleil, Natalie. 2009. "An impossible task?: Preventing disordered eating in the context of the current obesity panic." In *Biopolitics and the 'Obesity Edpidemic'*, edited by Jan Wright and Valerie Harwood. New York: Routledge.

Bernstein, Beth, and Matilda St. John. 2009. "The Roseanne Benedict Arnolds: How Fat women are betrayed by their celebrity icons." In *The Fat Studies Reader*, edited by Esther Rothblum and Sondra Solovay. New York: New York University Press.

Binkley, Sam. 2007. *Getting Loose: Lifestyle Consumption in the 1970s*. Durham, NC: Duke University Press.

Burgard, Deb. 2009. "What is 'healthy at every size'?" In *The Fat Studies Reader*, edited by Esther Rothblum and Sondra Solovay. New York: New York University Press.

Butler, Judith. 2004. "What is critique? An essay on Foucault's virtue." In *The Judith Butler Reader*, edited by Sara Salih and Judith Butler. Malden, MA: Blackwell Publishing.

Esposito, Roberto. 2008. *Bios: Biopolitics and Philosophy*. Translated by Timothy Campbell. Edited by Cary Wolfe. Vol. 4 in *Posthumanities*. Minneapolis, MN: University of Minnesota Press.

Esposito, Roberto. 2009. *Communitas: The Origin and Destiny of Community*. Translated by Timothy Campbell. Stanford, CA: Stanford University Press.

Fat Girls Guide to Living. 2009. *About FGG*. [cited 14/06/2011 2011]. Available from http://www.fatgirlsguidetoliving.com/about/.

Foucault, Michel. 1980. "Body/power." In *Power/Knowledge: Selected Interviews and Other Writings 1972–1977*, edited by Colin Gordon. New York: Pantheon Books.

Foucault, Michel. 1983. "On the genealogy of ethics: an overview of work in progress." In *Michel Foucault: Beyond Structuralism and Hermeneutics*, edited by Hubert Dreyfus and Paul Rabinow. Chicago, IL: University of Chicago Press.

Foucault, Michel. 1990. *The Care of the Self: The History of Sexuality Volume 3*. Translated by Robert Hurley. Harmondsworth: Penguin Books.

Foucault, Michel. 2000a. "The ethics of the concern for the self as a practice of freedom." In *Ethics: Subjectivity and Truth*, edited by Paul Rabinow. London: Penguin.

Foucault, Michel. 2000b. "Friendship as a way of life." In *Ethics: Subjectivity and Truth* edited by Paul Rabinow. London: Penguin Books.

Foucault, Michel. 2000c. "Sex, power, and the politics of identity." In *Ethics: Subjectivity and Truth*, edited by Paul Rabinow. London: Penguin.

Foucault, Michel. 2004. *Society Must Be Defended: Lectures at the Collège de France 1975–76*. Translated by David Macey. Edited by Arnold I. Davidson. London: Penguin.

Foucault, Michel. 2005. *The Hermeneutics of the Subject: Lectures at the Collège de France 1981–1982*. Translated by Graham Burchell. Edited by Arnold I. Davidson. New York: Picador.

Foucault, Michel. 2007. *Security, Territory, Population: Lectures at the Collège de France 1977–78*. Translated by Graham Burchell. Edited by Arnold I. Davidson. New York: Palgrave Macmillan.

Foucault, Michel. 2010. *The Government of Self and Others: Lectures at the Collège de France: 1982–1983*. Translated by Graham Burchell. Edited by Arnold I. Davidson. Basingstoke: Palgrave Macmillan.

Garner, Chastity. 2010. *The Curvy Girl's Guide to Style*. Createspace.

Grimshaw, Jean. 1993. "Practices of freedom." *Up against Foucault: Explorations of Some Tensions between Foucault and Feminism*. London: Routledge

Hardt, Michael, and Antonio Negri. 2001. *Empire*: Cambridge, MA: Harvard University Press.

Heyes, Cressida J. 2006. "Foucault goes to Weight Watchers." *Hypatia* 21 (2):126–149. doi: 10.1111/j.1527-2001.2006.tb01097.x.

Heyes, Cressida J. 2007. *Self Transformations: Foucault, Ethics, and Normalized Bodies*. New York: Oxford University Press.

Hoy, David Couzens. 2004. *Critical Resistance: From Poststructuralism to Post-Critique*. Cambridge, MA: MIT Press.

Hoy, David Couzens. 2007. "Response – reflections on critical resistance." *Foucault Studies* 3(November), 101–106.

Jarrell, Donna, and Ira Sukrungruang. 2003. *What Are You Looking At?: The First Fat Fiction Anthology*. Orlando, FL: Harcourt.

Koppelman, Susan. 2003. *The Strange History of Suzanne LaFleshe and Other Stories of Women and Fatness*. New York: Feminist Press at the City University of New York.

Lyons, Pat. 2009. "Prescription for harm: diet industry influence, public health policy, and the 'obesity epidemic'." In *The Fat Studies Reader*, edited by Esther Rothblum and Sondra Solovay. New York: New York University Press.

McNay, Lois. 1992. "The problems of the self in Foucault's ethics of the self." *Third Text* 6 (19):3–8.

McNay, Lois. 1994. *Foucault: A Critical Introduction*. Oxford: Polity Press.

McWhorter, Ladelle. 1999. *Bodies and Pleasures: Foucault and the Politics of Sexual Normalization*. Bloomington, IN: Indiana University Press.

Murray, Samantha. 2008. *The 'Fat' Female Body*. Basingstoke: Palgrave Macmillan.

Nealon, Jeffrey T. 2008. *Foucault Beyond Foucault: Power and Its Intensifications since 1984*. Stanford, CA: Stanford University Press.

Nietzsche, Friedrich. 2008. *The Gay Science*. Translated by Josefine Nauckhoff. Edited by Bernard Williams. Cambridge: Cambridge University Press.

O'Brien, Paul E., Wendy A. Brown, and John B. Dixon. 2005. "Obesity, weight loss and bariatric surgery." *Medical Journal of Australia* 183 (6):310–314.

Oksala, Johanna. 2005. *Foucault on Freedom*. Cambridge: Cambridge University Press.

Olssen, Mark. 2010. *Toward a Global Thin Community: Nietzsche, Foucault, and the Cosmopolitan Commitment*. Boulder, CO: Paradigm Publishers.

Ransom, John S. 1997. *Foucault's Discipline: The Politics of Subjectivity*. Durham, NC: Duke University Press.

Rochlitz, Rainer. 1992. "The aesthetics of existence: post-conventional morality and the theory of power in Michel Foucault." In *Michel Foucault: Philosopher*, edited by Timothy J. Armstrong. London: Routledge.

Royce, Tracy. 2009. "The shape of abuse: fat oppression as a form of violence against women." In *The Fat Studies Reader*, edited by Esther Rothblum and Sondra Solovay. New York: New York University Press.

Voss, Gretchen. 2010. "When fat comes back". *Women's Health Magazine*, May: 102.

Wann, Marilyn. 2009. "Foreword – Fat Studies: An invitation to Revolution." In *The Fat*

*Studies Reader*, edited by Esther Rothblum and Sondra Solovay. New York: New York University Press.

Webb, Cary. 2015. "Why I'm over the size acceptance movement or Hey, SA, What have you done for me lately?" *xoJane*, http://www.xojane.com/issues/why-im-over-the-size-acceptance-movement-or-hey-sa-what-have-you-done-for-me-lately.

Žižek, Slavoj. 1999. *The Ticklish Subject: the absent centre of political ontology*. London: Verso.

# Conclusion

## Style, solidarity and security

God helps those who help themselves.

Popular American pseudo-Bible verse

The aim of this study has been to examine the way lifestyle is used as a biopolitical *dispositif* or network that governs individual choice and population security. The example used to highlight this analysis was the obesity epidemic. I argued that it is not simply government health campaigns or medical authorities that target the everyday life of the individual, but rather a network of knowledges, practices, instruments, techniques and expert relationships surrounding the individual, some of which are engaged with voluntarily. Developing Foucault's notion of the *dispositif* I argued that the lifestyle network is composed of heterogeneous lines of power, knowledge and subjectivity that make the choices and bodies of individuals visible, knowable and governable. I examined the biopolitical use of the lifestyle network as an art of government that mobilizes freedom and security to govern the individual and population through norms. I also suggested that lifestyle is a device that makes some lives and choices visible as objects of care while others are the objects of exclusion.

A significant aspect of this analysis was the focus on free choice as a technique of governance *and* a practice of self-formation. Drawing together sociological analyses of Bourdieu and others with Foucault's ethics of the self, I identified an agonism in lifestyle consumer practices and subject formation. I suggested that the self is formed through agonistic struggle between different practices and choices that have competing underlying logics. That is, struggles occur between technologies of the self that use consumer practices to form the self-as-art *and* technologies of power that use consumer practices to form the self-as-responsible. In this sense consumer practices can both *style* and *secure* a particular form of life. However, the individual's enticement by, and voluntary embrace of, norms, disciplines, and subject-positions reveals the normative ambiguity of lifestyle governance. I argued that the lifestyle network of governance can be resisted in a number of ways, including: bodily resistance, counter-conduct, critique and caring relations with others. I used movements such as the fat acceptance movement to highlight these approaches to resistance. I argued that resistance within the lifestyle network is characterized by its everyday, endless and relational nature. Rather than a 'great refusal' or 'revolution', resistance to the biopolitics of lifestyle governance is subtle and tactical. However, I recognize that some may still object – 'what is wrong with using lifestyle to guide choice and behaviour change in relation to health?'

A case could be made that the lifestyle network of anti-obesity social marketing campaigns and smartphone apps is preferable to more disciplinary strategies. The Chinese government, for instance, has instituted mandatory calisthenics programmes to address public health concerns over sedentary lifestyles (Branigan 2010). In comparison, these complaints about lifestyle and neoliberal approaches to public health governance could be accused of being merely ideological. Or is there something substantially troubling about these developments?

To underscore the analysis made throughout this book, I will conclude with some broader implications. A major problem of the lifestyle network and neoliberal strategies of public health, such as social marketing, is that the public is reduced to the private. This reduction establishes a thin sociality that depoliticizes public health. Rather than addressing the structural and political factors inhibiting individuals and populations from living healthy and flourishing lives, these factors are depoliticized at the public level, while their effects are felt and suffered at the individual level. That is, the lifestyle rhetoric reinforces a belief that if you suffer from a chronic disease it is because of irresponsible choices. The lifestyle network makes the incidental choices and practices of individuals visible as objects of political concern, while masking the political nature of the structural factors that condition the choices and practices of individuals.

Choice and responsibility are mobilized in the lifestyle network to recode social welfare as an individual's capacity to provide their own welfare through private health insurance, gym membership and other care programmes. The dominance of responsibility in the lifestyle rhetoric of choice blurs the boundaries between the public and the private realm. These blurred boundaries are governed through various techniques of social marketing and smartphone apps to orchestrate the everyday life of the individual such that it 'atomizes our understanding of social relations, eroding collective values and intersubjective bonds of duty and care at all levels of society' (McNay 2009: 64). In this way, the biopolitics of lifestyle depoliticizes political and social relations by turning them into privatized relations of consumer choice that individuals are responsible for making. Rather than deep collective values and public goods, the network of lifestyle fragments social relations and thins out the idea public health, education and security. In this context, it is only a collection of self-governing individuals that remain.

The network of lifestyle alters the social landscape such that we do not all share the same fate. The individual that *appears* to be self-governing and responsible, that is the healthy subject, is enfolded into the secured population, while the irresponsible chooser with an unruly body or lacking the means to appear responsible is left exposed and vulnerable. As evidenced by the research used in the *Measure Up* social marketing campaign (see Chapter 3), racial and economic norms are used to predefine those who are cared for and those who are excluded from concern. For those who are the unable to adhere to the norms of the cared for, their inability to adhere serves to further reinforce their exclusion. This is despite the fact that social conditions preclude adherence. As McNay notes, the Aboriginal, migrant or poor 'are stigmatized as the "other" of the responsible, autonomous citizen' (2009: 64). Thus lifestyle network serves to bracket off hard and historically constituted issues of social determinants of health and subsequently depoliticizes public health by reducing health and welfare to individual choices and consumer behaviours.

These dynamics do not only occur in public health, but social policy more broadly. Earlier this year the Australian Prime Minister Tony Abbott described Aboriginal people living on ancestral lands in remote parts of Australia as making 'lifestyle choices [that] are not conducive to the kind of full participation in Australian society that everyone should have'. These comments came in the context of funding cuts to supply basic services to these communities. Prime Minister Abbott argued that taxpayers should not have to 'endlessly subsidise lifestyle choices' and the government is justified in withdrawing services and closing the communities (Medhora 2015). Arguing that these lifestyle choices were not conducive 'to full participation in Australian society', Prime Minister Abbott implied that Aboriginal people in remote communities choose not to participate in the nation's life, and are indeed irresponsible in a manner that indicates their not belonging. According to this logic, since this is their 'lifestyle choice', neither the government nor the Prime Minister can be held responsible if their lives deteriorate once services are cut.

Lifestyle and the thin sociality of neoliberal governmentality allow for blame and exclusion, but little room for care or solidarity. In the governance of public health, collective welfare and social security are recoded as an individual's abilities to secure themselves and not be a burden on others. Increasingly what is needed is not a self-styled security but solidarity with the excluded. As Foucault notes in one of his few clear normative statements regarding expectations of governments, 'we should expect our system of social security to free us from dangers and from situations that tend to debase or to subjugate us' (Foucault 2000b: 366). This book has used the obesity epidemic as an example of the way individuals are subjugated, but also to highlight the possibility for critique and resistance.

A related theme that has not been made explicit is solidarity. This overlaps with the argument put forward about caring relations that produce new norms for living together. Foucault expresses a form of solidarity that could be fruitfully explored in further projects in a statement at the UN in Geneva on the occasion of forming the International Committee against Piracy. Entitled 'Confronting Governments: Human Rights' Foucault opens his statement in saying 'We are just private individuals here, with no other grounds for speaking, or for speaking together, than a certain shared difficulty in enduring what is taking place' (Foucault 2000a: 474). This short and rich text offers many avenues to explore, particularly in relation to the idea of international citizenship and role of rights in Foucault's political philosophy (Whyte 2014). However, my interest here is the basis of solidarity that he alludes to. The mutual solidarity among those advocating change is based on a 'membership of the community of the governed' and 'shared difficulty in enduring' as witnesses to the suffering of those abused by power (Foucault 2000a: 474–475). The fact that we are all governed provides an initial binding. A subsidiary form of solidarity is as witnesses to those suffering under dominant powers and the obligation to never let that suffering 'be a silent residue of policy' but to use it as the ground to 'stand up and speak to those who hold power' (Foucault 2000a: 475). This idea of solidarity has potential in responding to the thin sociality of neoliberal governmentality and mechanisms of lifestyle. However, such an exploration cannot be entered into at this late stage.

This book explored the complex of knotted lines of power, knowledge and subjectivity that traverse the everyday lives of individuals in neoliberal societies. This complex of knots cannot be cut or untied, but continues to draw on and become entangled with new lines and threads. In tracing the lines that contribute to the emergence and use of the enabling network of lifestyle to govern choices and bodies, I did not seek to define a concept, but map the manner in which the individual's freedom, everyday life and subjectivity are problematized as an object of biopolitical governance. Rather than untangle the network of lifestyle, the aim of this book has been to bring into focus some of the frayed edges and loose points.

# References

Branigan, Tania. 2010. "Beijing workers shape up for return of compulsory exercises." *The Guardian*, 10 August.

Foucault, Michel. 2000a. "Confronting governments: human rights." In *Power: Essential Works of Foucault 1954–1984*, edited by James D. Faubion. London: Penguin.

Foucault, Michel. 2000b. "The risks of security." In *Power: Essential Works of Foucault 1954–1984*, edited by James D. Faubion. London: Penguin Books.

McNay, Lois. 2009. "Self as enterprise: dilemmas of control and resistance in Foucault's *The Birth of Biopolitics*." *Theory, Culture and Society* 26 (6):55–77.

Medhora, Shalailah. 2015. "Remote communities are 'lifestyle choices', says Tony Abbott." *The Guardian Australia*, March 10.

Whyte, Jessica. 2014. "Is revolution desirable?: Michel Foucault on revolution, neoliberalism and rights." In *Re-reading Foucault: On Law, Power and Rights*, edited by Ben Golder. New York: Routledge.

# Index